INVENTIONS

**20 Mechanical inventions
20 Methods or How To's
16 Theories and Thoughts**

by

FRED SANDERS

Bloomington, IN  Milton Keynes, UK

authorHOUSE™

AuthorHouse™
1663 Liberty Drive, Suite 200
Bloomington, IN 47403
www.authorhouse.com
Phone: 1-800-839-8640

AuthorHouse™ UK Ltd.
500 Avebury Boulevard
Central Milton Keynes, MK9 2BE
www.authorhouse.co.uk
Phone: 08001974150

First published by AuthorHouse 7/18/2006

ISBN: 1-4259-0964-7 (sc)
ISBN: 1-4259-0963-9 (dj)

Library of Congress Control Number: 2006903715

Printed in the United States of America
Bloomington, Indiana

This book is printed on acid-free paper.

This book is dedicated to:

My Father, Fred Sanders Jr.
For planting the seed of scientific interest
And to:
My Mother, Linda Laferty
For teaching me integrity and class
And to:
My sister, Lynn Fimple
For competing with me and
pushing me to better myself
And mostly to:
My Girlfriend, who wishes to remain nameless,
For offering so much time with research
and development of the sexual kind and
for being a good sport and a great woman!

INTRODUCTION

Most people find out that I'm giving out my inventions in a book and tell me that it's foolish. They say I should get patents and sell them to the highest bidder. That kind of greedy thinking is what has been holding me back for many years. Patents are expensive and many inventions require research and development that takes time that I can't fit into my work week. When I stand back and look at it, I see that greed is holding me back. I don't need that kind of money. As long as I can afford the basics, I would love to build inventions all day. I'm not doing this for money. I'll admit it, I would like to be recognized but that isn't my only motivation. I want all people to be more pleasant. I want to be happier and I want the people around me to be happier. I want to work less hours. I'm pulling for the blue collar worker. I want the human race to be able to continue to exist without having to struggle for our lives.

I have a special ability. I can solve the most complicated puzzles. I can do many of them in my head. I am a person who picked up a Rubik's cube and solved it using logic. Sometimes I can see a puzzle that others cannot see. I call myself an inventor. This is because I invent solutions to the problems I see.

Sometimes I have to convince others that a problem exists in the first place. What good is an invention that no one believes is needed? Many people turn their heads away from trouble. The same people who want to have children are the ones who are ignoring the damage to our environment. How do you think your children will get by when the planet can't provide enough food for everyone?

Sometimes I get very passionate about wanting to cure a problem. I hope my expressions will make you realize the size of the losses suffered as long as the problem exists.

Many of my inventions are just a better way to do the same old thing. I am a very practical person. I look for efficiency. Most of my inventions are new energy sources. If several of my new inventions were mounted to your car, your gas mileage could be greater than 80 miles to a gallon as of June 2005. By the time you read this, I could have a car getting 120 miles to a gallon. I'm always working on that. I say your car because my inventions can be fitted up to any vehicle, including a bicycle. I can recycle 50% of your pedal power. More people would ride bikes if the rides weren't so hard. (Even a bigger fuel savings!)

Some inventions are cures to problems I've dealt with, like my argument journal. It works miracles. Some inventions are cures to problems with our society. It's hard to call them inventions because there are no physical parts to them. In order for them to work, they will require public participation at some level, like taking your business away from the companies that sell you a known defective device. I think you would do that anyway if you were aware of it. I call these inventions "How To's" or "Methods".

The cures to the problems in our society might require you to handle a situation a little differently than you may have in the past. Many times, the losses are spread out among an entire community. When this happens, your personal loss may only be a nickel. That doesn't seem like enough to change your ways. It would be enough if you realized that most of the nickels lost by your neighbors will eventually come out of your pocket.

I want you to look at each of my inventions and ideas with an open mind. Remember that I'm trying to make all our lives more comfortable without injuring our environment.

First things first:

I am not a writer. I am a very slow typist. Writing this book is a tremendous challenge for me. I'm concerned that you may consider me arrogant. Some confidence can be taken the wrong way. I apologize in advance. I always have good intentions and I am not greedy. It is nearly impossible for me to prove this to you but I have to try.

I try to stay away from the technical terms because I know you don't need to know them and you probably won't remember them anyway. I don't think that you need to see technical terms to believe that I know what I am talking about. Some of my articles are confusing enough without them.

TABLE OF CONTENTS

THE HURRICANE TERMINATOR

Its name tells its purpose. It is hard to believe that this could be possible. I would rather build it and show you it in action but I can't afford to on my own. The only other way I can get you to believe in its function is to explain exactly how it works and what it will do. I will explain what a hurricane needs to keep it going and how the HT will starve it. To explain it I have to be sure that you are familiar with how a hurricane works. Then I can show you how I can build a machine that will disorganize the weather patterns that keep a cyclone going.

I'm not a meteorologist. I don't claim to know everything about a hurricane. A person only has to know the basics to know that we can cut off a hurricane's power source. Naturally, I will involve experts in several fields to assist in the design and operation of the HT.

How a hurricane works: Forget the swirling air near the eye-wall. Let's concentrate on the bigger picture. Warm high pressure air moves toward the lower pressure center of the storm. During the time it takes to travel there, it cools down. The bigger the storm, the more it can cool as it travels toward the center. The cooling affect is condensation. This means the air takes up less space and begins to weigh more per cubic foot. When the air takes up less space, it has less pressure against all those things around it.

Why does it cool down? It passes under cloud cover that blocks the sun and the rain passing through it will absorb heat from it. The rain is cool because it falls from a higher altitude where the air is cooler.

Why does the cloud cover remain above the warm air rushing below? The clouds at the edge of the storm are disbursing, but they are constantly being replenished by the thunderstorms in the eyewall and between the feeder bands. The expanding air in the eye is pushing up and outward on the thunderstorms causing a canopy affect. Some of the warm air finds an upward stream to carry moist ocean air over top of the existing clouds. As soon as it reaches the lower pressures it begins to form its own clouds along with more rain. This happens mostly in the eyewall but it can happen in smaller patches anywhere in the storm.

As you already know, thunderstorms occur when cool dense air comes in contact with warm air. Do you know why? The cool air is heavier then the warm air, thus pouring under the warm air and lifting it to higher altitudes where pressures are lower. The lower pressure allows more space for the molecules to spread out. This causes rapid cooling and rapid condensation. A cloud forms and rain falls. Thunderstorms occur on the eyewall where the cool dense air comes in contact with the warm air in the eye and between feeder bands where the cool air scoops the warm air upwards.

Finally, why is the air in the eye warm and why is the eye usually 20 to 30 miles in diameter? The air is warmed by the water below. I do not say land below because even the warmest land isn't enough to strengthen a hurricane, only sustain it on a sunny day. The air in the eye needs to be warmed by heat from the ocean below. The cool dense air from the eyewall rolls under the less dense and less heavy air in the eye. The heat from the ocean below warms this air causing it to expand and rise. The more it rises, the more it expands causing the funnel affect. The more heat transferred to the air, the faster and harder this air can push the thunderstorms outward. This is when the storm is strengthening. Over land, the heat and moisture is minimal and the thunderstorms thin out causing the storm to weaken.

The size of the eye is mostly due to the depth and pressure of our atmosphere. You will see what I mean by pouring just under 1 inch of water into a flat bottom round container. I used a large 4 serving yogurt bucket with about a 5 inch diameter. Make several fast tight circles with it then place it still on a table. Look at how the center stays lower than the sides. You might even see some of the bottom center of the cup with no water on it at all as the centripetal force keeps the rotating water against the walls. Now pour some of the water out and do the same thing. Notice how the eye is larger with a thinner layer of water.

The amount of heat transferred to the air and the weight of the cool air rushing in controls the relative size of the eye between 5 miles and 30 miles in diameter. The larger the difference of heat, the higher the rising air can travel, the larger the eye will be. The expansion of the ground level air as it rises the first 50 feet is critical to maintain the integrity of the hurricane. When a hurricane passes over land, it weakens. This is because the land can not transfer heat into the eye air fast enough to cause fast expansion and rapid rising.

The eye remains cloud free because the cool dense air cannot reach across the 10 to 15 mile radius to the center of the eye before it warms and rises. The inside edge of the eyewall is the lowest pressure part of the storm with the eye being almost equally as low. Some of the air from higher altitudes is able to lay low in the center of the eye. This air doesn't support cloud cover very well. In the experiment we did above, you can see how air can lay down in the center lower than the water on the sides during the time the water was swirling around. Some of this air is from the upper atmosphere where the temperature is much colder. This air can mix with some rising moist air to form hail. I have been pelted by one inch diameter ice balls doing about 40MPH sideways in Florida. Good times!

The swirling affect is a by-product of other greater forces. When heat is first cut off from the eye, the wind speeds will pick up for a moment as the eye shrinks. It gives the affect that the storm is strengthening when actually it is weakening. The swirl of air flow becomes a stabilizer. Air has inertia and takes time to move from one place to another. Just like in our cup, the swirl kept going for some time after we stopped giving it our energy. The swirl becomes the integrity. The swirl causes time delays on the other aspects of the storm, thus giving the storm the chance to feed later if it can't feed now. The storm can remain intact while passing over land.

I know that the wind does most of the damage and the wind is the swirl, but I also know that we are powerless to do anything to stop that effect. We have to cut off the power source to the storm and then sit back and wait for it to weaken. The hurricane Terminator will do that and one step more.

How will I cut off the heat source under the hurricane? I will use a blanket. It will be a sophisticated float with valves, water channels, pumps, sensors, computers, solar panels and bladders.

This giant float will be pulled around by mini submarines hanging underneath of it. The entire machine will be unmanned and controlled by satellite. It will be the largest mobile machine ever built. Luckily, it won't be the most expensive. Most of it will be basic floats made with life raft material. We will need a whole bunch of them to cover the greater half of the hurricanes eye. We won't need to make it 30 miles diameter because the first affect it will have is to shrink the eye. If we make it too small, the eye will attempt to relocate itself and end up dancing around the HT without having any affect. I will do more research to decide on a reasonable diameter to build it to. My first estimate is roughly 17 miles across. Keep in mind; the damage from one hurricane greatly exceeds the cost of building a 17 mile diameter machine. We have to build it. We can't keep suffering as hurricanes get worse as the earth's temperature rises.

How is a big life raft going to stay in the water with crazy winds whipping around? It will be large enough to act as a suction cup. The winds may be strong but they are not strong enough to lift 100 million gallons of water all at once.

Will the subs have enough power to pull it under the storm? We have one big advantage; the wind always blows toward the center of the storm. And yes, the subs will be very powerful and have an unlimited power source. The subs will be powered by the movement of the water around them and by solar panels.

With this machine being so big, how will it navigate boat and ship traffic? The machine will have the ability to sink itself in sections to allow boats to essentially pass right through it.

What is the one step more the HT will do? The HT will spray a fine cool water mist into the air above it causing its own small low pressure storm to form inside the hurricanes eye. This will shrink the size of the eye and cause a hiccup in the thunder storms in the lower eyewall. The denser air created will not be lifted by the cool air from around the eyewall thus disorganizing a critical part of the storm. The worst case scenario is that a handful of tornados will form near the center of the eye, but this is temporary and will happen out in the ocean. They will not make it to land.

Also, there has been evidence that hurricanes travel the warmest path in their general direction. The path of a hurricane is mostly controlled by the upper level winds but it is not completely controlled by them. We have watched several hurricanes ride up a coastline even though the upper winds will be pushing it inland. This is because the warm ocean can cause greater upward warm air streams then the land can. This causes the hurricane to pull toward the direction with the more heat from below. This means that the HT can be used to gently steer a hurricane out to the North Atlantic where it can dissipate above the cooler waters.

We can put it out in the Atlantic Ocean where it will cherry pick the storms as they form off the coast of Africa. There is a lot of distance between where hurricanes form and where we are. We will stay under the storm gently steering it away from land.

In the event where the upper winds cause it to migrate toward land regardless of our attempts to steer it away, we will just stay with the storm and attempt to keep it from forming an eye. With no eye, it's just a tropical depression.

One last note: It is predicted that with global warming, eventually we will experience what is known as a superstorm. This is a cyclone that never stops because it will always have the ability to feed off of the oceans heat. It will act in a similar fashion as hurricane Charley did in 2004. It will get strong in the water, cross some land, pass over more water allowing it to strengthen again, cross more land and then back to water, etc, etc. If we are not prepared with something like the HT before this happens then we will suffer non stop hurricane beatings all summer long, making the south eastern U.S. impossible to live in and possibly the rest of the country will suffer continuing bad weather enough that we might not be able to grow some crops. We could see our first superstorm very soon, possibly within the next 10 to 20 years. Imagine several storms that last 6 to

12 weeks each. I consider Charley to be our warning that they are coming soon.

Also, as I stated earlier, the greater difference of temperature between the ocean level air and the upper atmospheric air will cause a larger eye. Some have predicted that eventually the difference may reach a point where we will see storms with a huge eye where several layers of our atmosphere will be involved. In theory, the eye could be as much as 200 miles across. This kind of storm is only theoretical and may never happen. If it does, it will completely destroy everything in the eye. The HT will be useless against it, but don't worry; I personally don't believe this kind of storm is a possibility. Our atmospheric layers have too much of a pressure difference for a storm to engulf more than one layer at a time.

More information about this can be found on my website www. inventionsbyfred.com

MEN: HOW TO ACHIEVE 5 ORGASMS IN 5 MINUTES OR LESS

Each orgasm is more intense then the last! The guys I've told about this at work have come back to me the next day to tell me that they achieved 3 in a row on their first try. Their girlfriends were impressed enough that they would call several times every hour during the next day and bring them lunch each day of the following week

Here's how. Have sex as usual. When you feel it coming on, and you are sure there is no turning back, pull out. Don't use your hand for follow through. No hands means no sensitivity. You should still have a fairly normal orgasm. Wait for about 5 to 8 seconds after you are done, wipe it off, and put it back in. I usually add lube at this point. If you do, don't waste time. This only works if you stay on a strict time line.

Go ahead and return to good hard sex again. If you did it right, the climax feeling should return within 30 seconds. Do the same routine the second time, only this time, you want to wait only 3 to 5 seconds after you are done to get started again. Each time you go at it again, it will take a little longer then the last to get back to climax feeling. To compensate, I spend a little less time waiting after I am done ejaculating. The problem is that if you go back in too early, you will be sensitive. If you are a madman like me, you will learn to power through it to have the most incredible orgasm(s) life can offer. I usually scream!

I have been known to repeat this cycle 7 times. When I'm done I fall lifeless. You might not want to try this if you have a weak heart. The number of times I can do this is affected by whether or not I am using a sexual enhancement supplement and how much energy I have.

The time you spend waiting after ejaculating is the critical factor to pulling this off. If you wait too long, you won't be able to get back to climax feeling. If this happens, you can try and try for another hour without success. On the other hand, if you go back at it too early, you may be too sensitive to continue. When this happens, you can try to take another break but the sensitivity will stay with you for a while. It usually stays with me longer then needed to get back at it. The time waited after an orgasm is tricky. For me, it ranges from five to twelve seconds on my first time ejaculating and from two to six seconds on my second orgasm. Each orgasm after that requires less time then the last. By my fifth, I will only wait about two seconds.

Your time will have to be adjusted as you learn what works for you. If you wait 5 seconds and you are still sensitive then wait 8 seconds on the next time around. If that's not enough, try 12 seconds. If that is not enough then you may be moving around too much or pulling out too late. You may have to try several techniques before you master what works best for you. Try to keep the time as short as possible. When I wait more than 12 seconds it takes me much longer to get back to climax feeling. Usually I can get back to it within 30 seconds under normal circumstances. When I wait more than 12 seconds, it takes about 5 minutes to get back to it. When I wait too long, I never get back to it at all. I'll try for another 15 to 30 minutes before I give up. I call that a misfire.

I have told many men about my technique. Most of them have told me that they were successful on their first try that very same night. This is how I know that I am not a freak of nature.

A friend of mine said he didn't want to have several orgasms. He just wants to roll over and go to sleep when he is done. I explained to him that it only takes one more minute to have a second or third orgasm and each one is more intense than the last. I can understand his position. He is a little older and not looking for anything better than a basic reward. That's not me. I want the kind of reward that girls can have when they have 10 in a row. I may eventually figure out how to have 10 in a row myself. I wonder if I'll run out of semen and will I be shooting blanks?

I am in my mid thirties. The younger men I've told how to do this have all been successful at having at least two orgasms and they all agree that they increase in intensity each following time. Most tell me that they didn't use any energy drinks or enhancement medication at all, but the few who did had much more exciting stories the next day and a higher orgasm count. The older men don't seem to be motivated to try.

You know what you're getting into. I hope you use good judgment as to your capabilities. Please don't push yourself. Mixing energy drinks and other medications can be dangerous. I built up gradually to where I am now. I wouldn't dare just start out with full doses without knowing how they would affect me. Please consult a physician and use good judgment.

If you are using a condom, you might have to pleasure yourself after you are finished your first orgasm. I haven't figured out a way to get a condom on fast enough to stay within the strict time line in order to return to intercourse. Do not have unprotected sex just to experience multiple orgasms! I will not be responsible for your foolish misjudgment!

A REFRIGERATOR OVEN COMBO UNIT THAT YOU DON'T PLUG IN

You know that heat is a form of energy. You know that your refrigerator removes heat from the foods in it. What does it do with that heat? It dumps it behind the unit. This is wasted energy! Why does it take more energy from the wall socket to work? Why can't it run on the energy taken from the foods? It can! It doesn't have to be plugged in! I will build a unit that will store the heat energy until you need it later. The built in oven will use that stored energy to heat your foods without plugging it in to the wall.

This invention is very complicated to explain how it operates. You don't need to know every aspect of its internal operations. You only need to know that the refrigerator you have now takes energy from your wall socket and energy from the heat in your foods and dumps it in to the air for your house air conditioner to have to work a little harder to keep you comfortable.

This refrigerator technology will be incorporated in to your house air conditioner. The energy collected from the air in your house will power your outlets. If the air in your house doesn't have enough heat to run your outlets then the unit will use air from your attic or air from outside. Once made efficient enough, your house can be disconnected from your local power company.

This technology will be incorporated into your car. The roads become the biggest solar panel in the world when the sun hits them. Your car will scoop up the warmer air created by the sun hitting the roads and remove the heat from it. This heat will be used to make the rotating force that will propel you car. No more gasoline! This technology will reverse global warming.

Why isn't this done now? Efficiency. The existing methods used to turn heat into electric are too large or too slow to keep up with the demands we put on our devices. I've been developing motors that will be 100 times more efficient then anything used today. I have almost 10 different designs. The larger ones are the more productive. The problem I'm working on is making them small enough to fit inside your refrigerator along with a huge battery and still have space for an olive or grape. My first design will most likely have a remote battery and I'm kidding about only having room for a grape. Actually, my latest design takes up less space then the electric pump currently in your refrigerator. I still need more research and development on this version before it will have the temperature range needed to run a freezer.

The smaller the motor, the slower it removes heat from the air. We can compensate by giving the unit several smaller doors. The less warm air you let in, the less the motor has to do. I've also considered giving the inside walls of the unit a gel pack type layer to quickly absorb heat from the air now, giving the motor time to catch up later.

The house unit will have the advantage of having some yard, attic, or roof space for its components. The car unit will have the same issue as the refrigerator. We'll be lugging around some form of a mechanical or chemical battery needed to drive the car when the sun isn't out. Although, once it is made to be efficient enough, it won't matter if the sun is out or not. The air can enter my scoop at 35 degrees and I can still take energy from it then dump it out at 20 degrees. 15 degrees of heat is a lot when you are moving because you can scoop up 10 tons of air in a short drive.

Imagine how efficient it will be in the water. Water holds so much more heat then air does. A boat or sub could have an unlimited supply of energy. Although, it won't operate in near freezing water because if I take a few degrees of heat, the unit will freeze up solid. Air remains fluent until it is very cold. I will just have to design air units to be frost free like your freezer is now. No problem!

I know that this invention is worth a lot of money. Not necessarily to the seller of it but more to the lost sales of fuel and electric suppliers. This invention may cause me grief with the people who supply energy sources now. I hope everyone realizes that earth needs this, more then somebody needs to line their pockets with money. It would be smart for the energy suppliers to move their operations over to building this type of invention that will weaken their fuel or energy sales. It will keep their business alive after they stop selling energy. Energy is everywhere and it is abundant and should be free.

I feel that I am an expert at efficiency and energy flow. Energy is like water in our environment. Water evaporates in to the air, rains on to our land, runs through rivers and streams, and gathers in our lakes, ponds, and oceans. This is because this environment is a water conductor. Energy is just like water in the way it gathers here and there, evaporates into all objects, and runs through wires and other conductors. We just have to collect energy the way a water mill uses a waterfall. Mother Nature shows us how much energy is drifting around out there by destroying things with her wind storms. She is begging for us to gather this energy to use it to our advantage. The most efficient methods were used before we learned to burn fossil fuels. We used sails on our boats and wind to run our mills. We need to use today's technology and pick up right there, where we left off.

THE MEANING OF LIFE

The meaning of life: 1. is to feed your consciousness with pleasure. 2. The realization that you are ultimately responsible for the existence of your consciousness. 3. To make the changes you need to make to fulfill your goals of happiness, comfort, and pleasure for as long as is physically possible during your existence as self or race. 4. To affect others with your energy by passing on your personal influence. 5. Once your environment is a great place to exist, procreate.

I thought this was obvious. Sometimes we complicate things a little too much.

One more time in 10 simple steps:
1. Realize that you need to take care of all aspects of your life in order to feed your consciousness with more pleasure.
2. Make a decision. Decide to do the things needed to continue to exist.
3. Gather the tools, intelligence, materials, and wisdom needed to remain healthy and explore pleasures.
4. Locate new pleasures to replace the old ones.
5. Feed your needs continuously.
6. Consider the possible threats and develop methods to overcome them.
7. Put effort into your future; create an environment suitable for administering pleasure. Work to reach a point of comfort and happiness.
8. Feed your pleasures while trying not to do damage to yourself. Do not feed one pleasure too much. Use moderation.
9. Use your excess tools and materials to help others feed their pleasures.
10. Once you have made everything perfect for yourself and those around you, have children.

If you don't believe this then you will probably think I am obsessed with pleasuring myself. That is funny. However, you will eventually realize that I am right and I am sure of this.

I did not realize how many people did not know the meaning of life. I've come to a rude awakening. I realize that we need to work together toward the same goal. I realize that I can build an environment that works for me and any one of you can walk right up and destroy or steel it with very little effort. I now know that I need to build an environment that works for everyone. We will all need to reach happiness and comfort together. All of my inventions, ideas, and logic are backed by this intention. It has become my need and goal to make everyone happy. If I may be allowed, I will put all of my excess energy toward this goal and this will make me happy. I love to build inventions.

Notice how feeding your consciousness with pleasure is higher on the list than overcoming threats. Ultimately, if there is no possibility for more pleasure in your future, then you have no incentive to prepare for danger or fight for your life. You may realize this as you die. Realizing this may allow you to die peacefully.

HUMAN HEART ASSIST PUMP

My heart gives me some gentle pains sometimes. I can feel what is happening inside my chest. One or more of the muscles in or near my heart is clenching. It feels like a small cramp like I would get in my calf during soccer back in grade school. I know a cramp can last several minutes. If the right muscle were to cramp, that would be the end of me. I can handle the pain for as long as it takes for the cramp to go away. I just need to keep blood flowing to my brain until I'm through it.

The concept is simple. We mount a sheet like membrane just under the chest muscles with multiple channels for blood to flow horizontally through. Each end of the membrane will have one way valves causing the membrane to pull in blood when stretched and push blood out the other end when contracted. The natural breathing motions of your chest muscles will pump your blood for you in the event of heart failure. No motors or batteries to rely on, just good old fashion deep breaths.

We will give the unit several switches you can flip by pressing on a certain spot on your skin to control its amount of assist. I think it would be a huge benefit to healthy people who just don't want to over stress their hearts. Sometimes I feel like I could benefit from a little extra blood flow. I also believe that better blood flow will reduce the chance of getting cancer. Cancer seems to pop up in the low blood flow areas.

I have the feeling that my heart is my weakest link.

WHAT IS MORE IMPORTANT? WINNING THE RACE OR RACING THE RACE?

A while back, I found a place inside my girlfriend where I felt all the great feelings of hard sex, but I no longer needed to keep thrusting in and out. I've been experimenting with the feelings leading to climax. I've discovered that it's something like racing around a race track with no brakes. You give it as much throttle as you can give out of the gate. Once you are up to speed where you feel the big O heading your way, you have to back off the throttle for the up coming turn. If you noticed the turn too late, you'll go into it too fast and BOOM! Premature ejaculation! If you backed off the throttle early enough, then you can coast for a minute, scrub off some speed, and roll into the corner without crashing.

When I realize I'm not about to crash, I begin a slow steady pace just trying to carry my momentum. When I feel the road begin to straighten out, I can get back on the gas. I don't give full throttle right away while I'm still in the turn. A little faster and a little harder as time goes by until I'm back to full throttle.

I go back to watching for the next up coming turn. This next turn is a little different. I can go into this one a little faster and wait a little longer before I need to back off. I can stay on the throttle after I feel the big O coming on for just a little. After a few more turns, I'm going in to them at nearly top speed. I'm staying on the throttle right up to the turn. I realize now that I'm on the edge of the race track. From time to time, I feel my tires start to slide. I go back to giving just enough throttle to keep myself on the edge of the track. I can do this for long periods of time. Sometimes for 2 to 3 hours without a break. I stay right on the edge of a track where this turn seems to never end. After doing this for a long time, I get tired and I can see that my gas gauge is low; I'll deliberately speed up and cause myself to spin out.

Well, just lately, I've been getting into energy drinks during sex. It's not just like having a larger tank; it's also like having a larger motor. I've learned to get closer to the edge of the track and keep it there with just the right amount of throttle. Of course I've crashed many times before perfecting my technique. Now I've discovered something new. If I get close enough to the edge of the track where I'm half in the dirt and half on the rumble strips (I'm just about screaming), I can stop thrusting and just push in steady and hard. I can keep the crazy off the chain feeling going for a few minutes without moving at all! I just keep the pressure on steady and it keeps me gritting my teeth and clenching my muscles and breathing so hard that it sounds like I'm hyperventilating. I'm still not moving! I haven't budged in minutes! After a few minutes, I find myself yelling AAH! AAH! AAH! This will happen for as long as 30 seconds or so. I can't seem to find my stop watch when I need it the most. Suddenly BAMM! I pull out and I'm losing it all over the place. At this point, I do like described in the multiple orgasm article. I don't use my hand at all. I don't need it. The advantage to not being sensitive afterwards is getting right back on the edge of the race track.

If I play my cards right, I can do this pattern several times. I can stay in the climax feeling up to 30 minutes. I might ejaculate 3 to 7 times. My girlfriend has been there for me while I climaxed over and over without a break. It's all about stamina.

When I'm by myself, I can hold out longer and have more orgasms. I know the only difference is that I'm not spending as much energy when I'm by myself. I can hold out longer before I have to fall lifeless for about 20 to 30 minutes. I don't save enough energy for clean up and I don't know why.

For you to master hanging on to climax feeling, you will need to know how I get there. You've read the above saga so you have the basic idea, now you just need some pointers. In the beginning of the above article I say you come out of the gate at full throttle, but that was just to make a comparison to racing. The truth is that you should start up slowly. Let it take 10 minutes or more to build up to full speed. Getting past the first turn without crashing is just a matter of noticing it coming early enough to slow down or stop. It's common knowledge to think about sports or weather to take your attention away from the great feeling. Although, that can backfire if your meteorologist is sexy. I don't do that. I prefer to slow down. My girl likes it slow.

Here is the part you will have to get past mentally. Almost as soon as you slow down, the climax feeling seems to go away completely. So naturally, you want to head right back to it at full speed. Your second time there can surprise you. Start back at it slowly. You have to pay attention to it more this time. Be prepared to slow down or even stop if needed. You can repeat this cycle about 4 or 5 times before you may notice that it is starting to become easier to get close to climax without crashing. You will have to push the envelope 4 or 5 more times before you will notice that you don't have to back off much any more. You just slow down a little bit. Remember that you want to stay in the climax feeling without going overboard.

Still, you will have to back off 4 or 5 more times before you reach a point where you no longer have to back off. At this point, you can stay at ¾ throttle continuously right on the edge of the track in the climax feeling for long periods of time.

The last thing to know is that you will have to punch it up to full throttle for just a few seconds every once in a while, in order to push the climax feeling higher. I have taken my climax to the level where the feeling is so insane that we get so loud that I begin to expect the cops at my door after a while. I can go until I no longer need to thrust in and out and I have such an intense orgasm that I have to scream. We laugh about it afterward. Something like this never gets old.

I know I couldn't have figured out all of this without the help of energy drinks, erection enhancement supplements, and a goddess for a girlfriend. Thanks, Honey Bunny!

MY THEORY BEYOND THE BIG BOOM

If you don't have an open scientific mind and some chemistry background, then don't bother to try to comprehend this!

I have reviewed the data provided by the scientists who have been trying to piece together the building blocks that make up our existence. I have reviewed their testing methods and I have found flaws in both the tests and the math used in the results of those tests. I believe the currant accepted theories are off base by about 33%. This has caused confusion amongst today's scientists and has lead them to believe that the laws of physics break down at the atomic and sub atomic levels.

My theory allows for the laws of physics to remain intact at the atomic and subatomic levels. I call my theory the Sanders Theory. What is it going to take for my theory to gain acceptance? Proof. This can be difficult to achieve because we don't have the ability to see subatomic particles. The current accepted theory has no visual proof for the same reason. So what is our proof made up of? It is made up of data from several tests combined with the parameters of the basic laws of physics and crunched with mathematical equations.

The first thing to consider is that we wrote the laws of physics based on what we can see. That is how we thought the Earth was flat. I am an inventor. Inventions require an incredible imagination. You know by my inventions that I am good with this ability. Well, I'm sorry to drop this on you, but my imagination allows for our laws of physics to be a sphere where other scientists believe they are flat!

The second thing to consider is that we didn't consider all the forces applied after running the tests that we now hold as our proof. Forgetting to factor in one small force will lead to all of our numbers being off by a small percentage. In most cases, a small percentage off will remain that same small percentage off through all of the math. This case is different then most. In order to reach our present beliefs, we had to use math on top of math on top of math, etc. If you trusted the results from a previous scientist and used his calculations, you may have been using incorrect data. In my theory, all scientists have been using incorrect weighing methods and measurements for centuries. These methods have been accepted by all people for your entire life and therefore you haven't had a reason to double check to make sure that the basic units for measuring weight are correct.

What does this mean? It means any theory that has weight measurement in the equation may be off by a small percentage. In fact, nearly all of the theories believed today have weight in their equations. As I stated earlier, math on top of math makes a small percentage become a large percentage.

Why did I feel the need to double check the math and weighing methods? That shouldn't be the next question. It should be; why didn't our present scientists' double check the math and weighing methods the first time they ran in to equations that didn't work out? Do they lack logic? Are they blinded by beliefs? I thought religion caused people to believe things without proof. Has science become their religion?

I know that the math always works out. So I ask, why doesn't your math work out? We have to treat this like it is a technical malfunction. We have to go backwards, double checking all of our equations, measurements, and methods of testing until we find the flaw in the math. At the same time, we have to have an incredible imagination.

Every time we double check a testing method, we have to realize how a test is compiled. We have to remember that a test starts in the imagination. Why would we run a test if we can't imagine more than one outcome? The only reason for running a test is because our mathematical equation has two or more variables. If it only had one, we wouldn't need to run a test. We could just crunch numbers to get the answer. We were and still are relying on the imaginations of people who did not have our quality of education!

Here is how it works; you set up the test using known information and you imagine the possible outcomes for all the variables. You run the test and neatly place the results in the parameters you imagined, even if the results land just outside the possibilities you imagined. Since all tests have always been performed in this fashion since we developed intelligence, we have to realize that the original data is just as shaky as the data that doesn't compute today. A great diagnostician works from the middle out. In this case, we need to crunch the numbers forward until something doesn't compute, then backtrack until the numbers don't compute, reformulate, and go forward again. The day we lose sight of this method is the day we stall the engine. How far will we glide with the engine stalled?

I asked myself; what parts of their equations are still not fully figured out? Magnetism doesn't fit neatly into most of today's theories. In fact, it is the one factor that is throwing off most of the equations. We have to study magnetism in order to find the rest of our answers. I did!

I studied magnetism until I discovered that it is a small force that we didn't factor in to our weighing methods. I believe we miscalculated the weight of the proton, neutron, and electron. I saw no evidence that magnetism was factored in and therefore, we may have wrong measurements by a small percentage.

The poor weight measurement is not my basis of the Sanders Theory. It's just the motivation to search for a new theory. I wanted to find a theory that allowed for the laws of physics to remain intact at the subatomic level and also explain magnetism with absolute certainty. I did. I came up with a single possibility that works in all equations I've done so far. In order for my theory to gain acceptance, I have to prove something that can't be proven with the present accepted theories.

This is difficult to do because the bar has been set very high. A person needs a lot of funding to have access to the kind of research equipment needed to find proof. That person needs to dedicate themselves to this cause. So for now, my theory just remains an unproven theory while I handle other aspects of my life.

Before I explain my theory, I want to be sure that you know the difference between my theory and the existing accepted theory. It is believed that an electron has a negative electrical charge. It is believed that a proton has a positive electrical charge that pulls the electrons in toward the proton. It is believed that the electron travels inside a shell but does not fall all the way to the surface of the proton.

The Sanders Theory: There are only 2 subatomic particles that make up the significant parts of atoms. Both have a gravitational pull and inertia related to their mass but only one, the electron, has a repelling force. Protons and neutrons are the same and are not electrically charged. The difference between a proton and a neutron is that one has an electron stuck to its surface and the other doesn't.

There are only 2 opposing forces at work. Electrical charges and magnetism are both side affects of gravity and the repelling force of electrons.

Only the Electron has a substantial repelling force. The repelling force between two electrons is greater than the gravitational force between two neutrons at the same distance when extremely close together and less when further apart. This is due to size. Gravity pulls an electron in to the surface of the neutron and that neutron becomes a proton. A second electron will be pulled in by that protons gravity but it will only descend until it reaches the equal repelling force from the first electron on the surface of the proton. This second electron will settle at the closest point to the proton on the far side from the first electron. The assembly will move as one unit when cool and the nucleus will vibrate/spin at a different pace when warm. Gravity will hold it together and the electrical repelling charge will hold it apart.

When cold, all three parts will spin at the speed of the proton because it has the most inertia. When more heat is applied, the locked, closest electron to the nucleus will break away from it's locked in position to travel a path no longer across from that of the one on the surface of the proton. This will cause the appearance that the electrons are moving further from center as if to move to a higher shell.

The gravitational pull of a neutron covers more volume then the repelling field of the electron. Also, gravity is like light. Overlapping causes amplification. The repelling fields of the electron act the same way compressed air works. They only overlap equivalent to the pressure around them. I want you to think of each electron's intense repelling field as a perfectly round balloon. Two balloons can be presses toward each other and as more pressure is applied, the more surface area of the balloons will contact each other causing more resistance to them approaching each other.

Electrons have no friction to each other and they have very little inertia. This will cause them to appear to pop in and out as the nucleus spins and the electron on the surface passes that side of the atom. Since electrons each have roughly the same size intense repelling fields, the shell shape of different size atoms are similar.

The larger the gravitational field, the more electrons will get caught in it. Atoms will move around as a solid with properties similar to a bunch of balloons tied together at the center. The outer most electrons will have gaps between the repelling fields but still within the gravitational field. Any two atoms with crossing gravitational fields will cling together. This is the molecular bond.

The nucleuses will pull as close together as possible without letting any two electrons cross each others intense repelling fields. The semi-loose electrons will be pushed out to the far sides of the molecule causing their repelling fields to reject bonding with other atoms or molecules. It would be like tying the bunch of balloons together tight near the center and loosely on the outer layer. If you can make some space between balloons on any two bunches, enough for the two bunches to be closer than the diameter of one balloon from one another, then the nucleuses will achieve a gravitational bond. If one bunch has enough space between the outer layer balloons to allow two other bunches in close enough, it will bond with both as in a water molecule.

My theory states that there is no magnetism. The attracting affect is actually gravity. The repelling affect of two magnets is actually the electrons electrical repelling force. The two forces appear to be about equal because all atoms and molecules are about equally balanced. Each atom's gravity pulls enough electrons to make a near balance between push and pull. Magnetism is the result of the molecules cooling to becoming locked in a direction where there is more pulling force on one side, the same side, of each of them and this causes the greater repelling force to become held to the other side.

It would be logical to ask "Why doesn't a magnet repel everything from that side?" The reason is that all atoms have parts that are constantly in motion in a low resistance environment. The repelling and pulling parts simply rotate in such a fashion that they balance out other forces applied to them. The only molecules that can't balance out the forces applied to them are found in magnetic substances.

We are looking at it backwards. It's not the magnets that are doing something impressive; it's all nonmagnetic substances that have the ability to rotate their loose particles until they find balance. Only magnets don't have this ability. Did you know that human skin has magnetic properties?

I believe the laws of physics remain constant at the sub atomic level. This would help explain how all neutrons are relatively the same. I have to start my explanation at the end in order to have you understand the beginning.

If the laws of physics remain constant and large and small are infinite, then all things should be recurring in similarity, possibly in equal size intervals. The laws of physics should force this result. This is because a low friction environment such as space will always be an even playing field on any size scale.

Here is my reasoning: Once gravity reaches a certain point, the body emitting that gravity will collapse in on itself. This should happen at about the same relative amount anywhere in space. Each black hole should form at roughly the same size. Each of our laws of physics has its break points at certain intervals. This will cause equality among the resulting remains.

The electron's repelling force can hold X amount of pressure. The pressure is created by the gravity of neighboring neutrons. The pressure should be about the same each time a body collapses. The resulting bodies should occur at roughly the same size each. This is how I believe our neutrons were formed. The smaller debris collected until it reached a break point where it collapsed and solidified under its own weight. This debris might be electrons that lost their electrical charge.

I can't seem to find a word to describe a repelling field the way gravity describes the pulling field. Do you like "resisity"? The trick will be to use this knowledge to amplify the resisity field to create antigravity. I have a few ideas.

If the resisity breakpoint is the same for all electrons and the recurring effect is true, then we must study the large in order to understand the small. Comparisons of similar circumstances at different levels of size will help our understanding. For example: compare our solar system to a single atom, our galaxy to the tiny little spot of dull light glowing from inside a flame, and our universe to the entire flame.

If time were infinite, inertia were constant, and the laws of physics do not break down at any level, it would make sense that time is related to size. Our electric moves at a constant rate, therefore, our energy moves through all things familiar to us at a rate we compare to important occurrences to us at our size. Again, the same sentence using other terms, the speed of energy flow at each particular size, is the control of time at that size.

If you don't understand so far then try this. A large creature will view time at a slower rate than a small creature. A minute is shorter for me then it is for my cat. Here is my last example: If a creature were as large as 10 of our solar systems, it would perceive the speed of light the same way we perceive the speed of heat conduction through a piece of metal. Time is a perception based on the speed of energy movements.

My theory means that time for us is super slow motion to all of the life forms that form, evolve, and die on a much smaller scale. The life forms are so small that entire civilizations could live, evolve, and die off on a fire's ash in what seems to be a millionth of a second to you. When that ash cools off it will no longer be able to support life. It will combine with other solids the same way all material in our galaxy will end up in a black hole. Consider that a full black hole is like a neutron. Maybe it would take an entire galaxy to complete a black hole.

Over time, all the galaxies in our universe would become the atoms in the next larger realm. My theory is that each realm is similar to the last in design but much larger because the laws of physics in a non friction environment such as space should have a repeating effect as long as time is eternal.

Our existence as life on earth will amount to less than 1 second to the life forms living in the next larger realm. If there were tiny life forms living on the ash of your candle's flame, they would not be capable of affecting our realm. Just the same, we are not capable of affecting the next larger realm.

The life inside a fire is an example of size relations more than it is a theory of mine. A more realistic theory is that energy is passing through matter keeping it spread out. As matter relieves its energy, it condenses in patterns. The same kind of patterns we see in atoms is what I see in our universe's future.

Life occurs when a condensing pattern gains the consciousness that it has the ability to alter this pattern. This occurs after the organized pattern can tell the difference between good and bad. The most likely first sense is taste or something like taste. The organized pattern will take in individual particles to decide if they will benefit or harm the organization. Life originates when a feeling comes back from the organization and that feeling can last until something solidifies or a memory is stored.

I've stated many thoughts in this article. I know that many of them will be rejected by many people. The people who reject these ideas may not be able to imagine life where electrical charges and magnetism are side effects of gravity and resisity in a non-friction environment. It all comes down to belief. If you believe something that you have poor proof of or no proof of, then you will not be able to open your mind to other possibilities. Even if you don't believe a word I say, you have to admit that this is very interesting material. Imagining possibilities is fun.

I don't believe that these theories of mine are accurate. I believe that they are closer to accurate than any preexisting theory. So far, I've been able to explain every occurrence with the Sanders Theory, even how the pseudo forces (magnetism, electrical charges) are just side affects of gravity and resisity in an environment where the laws of physics do not break down at any level.

THIRD BRAKE LIGHT UPGRADE

This will save lives and reduce wasted resources. Whether the car in front of you stops slowly or stops hard the brake lights are the same. It's time we put a tiny pendulum inside the brake light housing on a position sensor that will light a row of red diodes in succession according to how hard we are breaking. The diodes will start at the center and spread out further the harder you stop.

If all cars are given the same size pendulum and row of lights, then all cars will give the same readout according to how hard they are stopping. We will quickly learn a comparison and be better prepared for a hard stop when we didn't expect it. This can be made so cheap, it's nothing compared to the cost of the damage it will save. One crunched front end can cost more than 1,000 brake light units. I'm not including the damage to the rear end of the other car or the injured people. The same pendulum could light a row of green diodes to show the amount of acceleration. This may not help stop accidents but it can alleviate a little traffic build-up where people will wait to accelerate because they aren't sure how hard the car in front of them is accelerating. More people will get through during the time the light is green, thus reducing how far traffic will be backed up at the red light.

Do you believe your teeth and gums are the only thing you can put into shape with braces? Your entire body can be shaped in the same way. It won't happen over night.

A child sucks his thumb too long and gives himself buck teeth. A parent stands his child too early and gives him bow legs. A girl who puts her long hair behind her ears for several years in a row will slowly push her ears outward. Grip handle bars all day every day for your teenage years and you will end up with large hands. I did! A tribe in Africa has a custom where they tie tight bands around their children's foreheads in order to cause the top of the skull to grow upwards instead of outwards.

It's not just your bones. You can build up or squeeze away tissue or fat from certain areas. I believe this effect is temporary. Flexing the bones is a little more permanent.

If you want to have a thin waist then you should wear a corset or girdle day in and day out for a while. I can't say for certain if a penis pump can have any long term effect because you are not shaping a bone. (No pun intended) A girl can sleep with a pillow between her legs to give herself wider hips. (Wide hips make your waist look thinner!)

If you want thicker skin, you have to treat it the same way you treat the pad of your foot. (That is how my hands grew to be when I did construction.) If you want thicker, stronger finger nails, just pull on them in the same direction that they would be pulled on if you were an animal and you were climbing a tree.

Your nails and teeth are the last little bit of what was once an exoskeleton before we evolved. The same goes for the hooves of a horse and the horn on a rhino. We were all like bugs a few hundred million years ago. The truth is that we shaped our bodies in the way that worked best for our survival over the centuries. Now that fighting for our survival is not as much of an issue, we can continue to shape our bodies any way we please.

Your bones are most flexible when you are very healthy and active. This is why young people can handle bouncing around so much and the people who "settle down" become so brittle. Bone hardening is actually part of the healing process. Every time you get sick for any length of time that keeps you in bed, your bones harden a little more. Lounging around causes poor blood circulation and that leads to most parts of your body becoming malnourished. Low blood flow results in slow breathing and less oxygen makes it to your muscles. This will give you a tired all the time feeling. If you are experiencing this feeling, you are slowly withering away!

As for the rest of us, if you remain healthy, you can keep your bones soft enough to push them into shape without breaking them. Be careful not to cut off circulation with tight cinches or wraps. If you don't like the position of a few of your teeth, you can grab a hold of them and gently push them where you want them for at least 15 minutes every day. To make this work, you have to have some self control in the chewing department. When your teeth begin moving, the chewing faces will no longer line up. A dentist will grind them in a way that will make them comfortable. You can just chew on the other side for the time needed to get your teeth into position. Once they are in position, you can go to a dentist and ask to have your teeth contoured.

Any part of your body can be reshaped as you see fit. Thinking back almost 30 years, I had a problem as a child where my feet pointed inward. I didn't feel like anything was wrong until somebody told me that it looked funny. From that point on, I deliberately turned my feet outward just a little as I walked. I only had to concentrate on it for a few months. Since then, nobody has told me that my legs or feet look weird. It is much easier to shape yourself when you are young.

I don't know when it's too late to try to flex your shape. I have to assume that it will be different for each person based on health.

Consider doing this before you see a marriage counselor or before you send your child to boot camp.

Do you argue or fight with anyone you love? Do you have a once in a while disagreement that leads to walking away with feelings like your not understood? These feelings can be tucked away neatly and forgotten. They can build up without being noticed. Eventually, they will come crashing out of the closet during a simple argument about whose turn it is to do the dishes. Even during this episode, you still won't be able to explain yourself well enough to be completely understood and even with a good ending, you will still be tucking feelings away for a later day.

I found myself trying to find the right time to bring up the subject I knew would lead to an argument. I would avoid talking about those sensitive subjects if I thought it would lead to a heated battle. Every time we argued, we would throw in that other little detail that was related but not helpful to the arguments primary subject. Next thing I knew, we were so far off the original subject, I knew we weren't fixing the problem.

I found the cure on a note. The problem got fixed if it was written down on paper. Why? What is the difference? When you write down your problem, you get to finish your complete thought. When the other person is reading the note, they might want to stop to call you names or explain their actions or blame the other guy. This is where the argument steers away from the original subject. If you are not there, they can't cut you off before you are done your complete thought. They have to keep reading the rest of the note until finished. This is where you get to say "I know why you did what you did and I'm not holding it against you. I just want to fix the problem."

I want you to know that arguing on paper has made my life much better. I've always had a good life. Now I have a great life.

I started a journal with my ex-girlfriend. Within one week, we went from sleeping separately to being affectionate regularly. Within the first month, we worked out all of the obvious problems we had.

We were at a point where we were writing "All is well today. Did you have a good day? I love you, Fred"

We drug it out for about 6 months. During this time, we did catch several more problems that were not obvious. They were no trouble to take care of. We quit using the journal because we had no reason to argue. For the next 6 months, we loved each other more than ever. Then a few troubles came up. We were back to arguing again. We agreed to use the journal again. It cured the simple troubles and it showed us that we weren't right for each other. I love her still. I will always love her. Now I know that we are not perfect for each other. We agreed to move on.

My present girl and I agreed to keep a similar journal. We don't argue anymore. We love each other very deeply. I know I could not love her as much when I let the little troubles fester. Again, we quit using the journal after about 6 months. Why write "I love you" every day? It's more fun to say it face to face. It's been almost a year since we stopped using it. It's been a wonderful year with no arguments at all. I can tell that we have a few new feelings that we aren't telling each other. I suggested starting it up again. She was reluctant. She doesn't want to go back to writing "I love you" every day.

In order for the journal to succeed, you have to agree on a few basic rules. You have to write something every day for the first month or two, just until the major problems are worked out. You must allow the other person to have complete privacy. You can not read their entry before it is finished. You might find yourselves making several entries in one day. This is great. Work out those problems faster. I don't know if a three person journal can work. You have to let the personal feelings flow. I'm not so sure I could do that knowing that two others will be reading it.

You have to come to terms with the possibility that the journal might show you that you are not right for each other. It's better to find out now then to find out after several more years of tough love. The journal can make somebody cry. This is how we learn. I could hear my ex-girl crying while typing her response to my entry. By the end of the night we were in each others arms.

My present Girlfriend found out that I still love my ex-girlfriend by reading this article. She confronted me because she doesn't understand how. I told her that I still love most of my ex-girlfriends. To help her understand it better, I explained that if some outside reason caused us to break up right now, would she stop loving me a few years from now? If she were asked 5 years after I was gone if she still loved me, how would she respond? She responded with "I will love you forever." My response… "Touché!"

YOU KNOW THE RISKS.

A friend of mine was sued because the sidewalk in front of his house was being lifted by a tree's root and a neighbor hurt themselves on it. I don't know how we can sit back and let this happen. This means I may have to take time out of my life to deal with lawyers and insurance companies or even have to pay my very hard earned money to someone (and their lawyer) who never learned how to negotiate a simple bump in the sidewalk. If you are his neighbor then you knew the sidewalk was heaving before you left your house. You knew the risks. I made it a point to learn how to handle rough terrain and now I might be punished because I live near someone who didn't!

We've all heard of the hot coffee incident. You know that coffee is hot enough to burn you. You've watched it percolating. It's obvious this is not right! I'm not saying you shouldn't get your medical bills paid. I'm saying you don't deserve pain and suffering money when you know the risks, especially when you have the option to avoid the risk. Coffee isn't even good for you! If I'm playing catch with a burning charcoal briquette and I catch it with my eye, can I sue the guy who sold it to me? He didn't tell me it would get that hot.

A different kind of insurance policy system would take care of this and reduce the amount we all pay. Every person should have to take a policy out on themselves that covered them out of the car the same way as in a car. Do you sue your insurance company for pain and suffering when you hit a pole? How about the telephone company? You just make a claim. Nobody would be allowed to sue other people just because they were on their property when they became a klutz. You know that this is incentive for people to set each other up. Also, if you could only make a claim to your own policy then we would be leaving out the lawyers. There would be the big savings on premiums.

You could still sue a person if they had a direct deliberate part in causing the accident. I still don't consider a piece of heaving pavement to be my friends fault. The neighbor knew it was like that and could have handled it differently.

I'm afraid to let the neighborhood kids come over to play with our boy. Will I lose everything if one of them got hurt? I hope my sidewalk stays flat. I actually fear that I may become a victim of a poorly designed system. Is it just easier to justify insuring property over insuring people or is it driven by greed?

A system where you carry a policy for your self would make the people who have the most claims have to pay the largest premiums. The bigger the klutz, the more they pay. A person might say "That's not fair! A person won't be able to afford insurance after several claims." Tell me why you and I should pay it for them? Does Phys Ed require agility requirements before giving a passing grade? Should it? Will an A lower my premium?

If I begin a crusade to put an end to people extorting money from other people and businesses, would you be willing to sign a petition and lose your right to "get rich quick" by extortion? If we control this problem, we will all pay less for our insurance. We will no longer have to pay for the legal battle.

I feel it takes an evil person to believe that extortion is OK. Nobody should profit from an accident.

Another way to control this problem would be to start a new kind of insurance that would insure that in the event a policy holder is being sued; several good attorneys would defend them without charge and without ever offering a settlement to the plaintiff. The chance of someone getting pain and suffering money would be reduced substantially and in turn, the suing attorneys would suggest not pursuing, just filing a claim.

THE MOMENTUM RECYCLER

This is my oldest invention. Twenty years ago, I invented a large version of that little car you had as a kid where you would push it forward several times to spin a little weight inside faster and faster. The weights inertia would keep the wheels of the car spinning with enough torque to climb over small hills or just drive slowly down the sidewalk. I made a larger version for my bicycle. My daily ride was a couple miles of steep hills. I could spin up a flywheel on my way down a hill and use its inertia to make it up the next hill. Since then, the idea has evolved several times. The spinning weight is just a form of a battery where energy is stored until needed later. The problem with that is the weight would spin down after a few minutes if left unused. The battery would discharge.

A few years back, a major car company produced an economy car with a similar spinning weight. Foolish! Didn't the designer ever play with the toy when he was a kid? It's a poor concept. The battery is too heavy to be practical. Maybe that is why we don't see too many of them on the road today. I have a better plan! We can use a battery that weighs less and doesn't discharge when sitting still. The battery can be a spring mounted directly to the drive axle that already exists in your car or truck. In fact, we can put one on each of your axles. The spring will be of the coil type and be just a little larger than the axle. The brake line to the caliper will T off to a second caliper mounted to a small rotor on the input side of the spring. As you press the brake pedal, the second caliper will clamp down on the small rotor and the rotor will begin to twist the coil spring. You will still have complete modulation for smooth stopping. When the coil spring is completely wound up, the rotor will begin to spin at the same rate as your stock rotor. You will have the increased stopping power of additional brake assemblies. Your brakes will last much longer because some of the energy that would have been turned to heat by your break pads will be in the wound up spring. The next time you press the accelerator, the spring will release the tension back into the axle, helping to get your car up to speed. When you let off, the spring will stop pushing the axle and allow the axle to freewheel. The remaining spring tension will be there for next time you press the gas pedal. You won't have to press the pedal as far to get up to speed. In stop and go traffic, you won't need the engines power at all. The spring can do that for you.

This invention will increase your city gas mileage and increase the life of your brakes. When it is made efficient enough, your city gas mileage will be much better than your highway mileage. This is because you are pushing wind on the highway and not as much in the city. The unit can be made generic enough to fit any vehicle or it can be tailored to work at maximum capacity for the size of the car. I would put a small unit on each wheel rather than one large unit in the middle.

I think this is the most important invention to make right now. We must take a fair size step toward saving our environment right now before global warming causes more destruction then we can handle. We have to reduce the amount of fossil fuels we burn. This device will cost about $300 to make and will save the average driver about half of the expenses between gas and maintenance from that point on. The average person will save $300 on gas alone within a few months and will be protecting the environment by not burning as much fossil fuels. Your same old car will **accelerate better then ever** with my invention attached. We must all do this as soon as we can. We all win with this invention. I fear this invention will bring me grief because oil companies will see future sales deteriorate.

This invention combined with a solar panel and an electric drive motor would be enough to drive your car around without starting the engine. The electric drive motor wouldn't need to be big at all. It would only have to maintain your speed. The springs would provide acceleration. Think of the benefit of being able to drive your car to a gas station after you run out of gas. Most cars are hard to steer with the engine off. I can imagine a car where we would start the engine for the purpose of air conditioning, heat, power steering, and electrical power, but we won't need to rev the engine. Imagine how long a tank of gas would last if the engine only idled.

THE MUSCLES IN YOUR EYES

I'm not getting into this subject too deep. It's obvious that muscles in your eyes allow you to focus at different distances. If you can't see well far but you can see well near, then this proves your eyes work well in the other areas. You just need to teach the muscles in your eyes to stretch further or pull harder in order to make you be able to focus on a larger range.

I've always had great eyesight. About 5 years ago, a child read a sign a quarter mile up the road before I could focus on it. I realized then that my eyesight wasn't as good as it used to be. Logically, I figured I would practice focusing at distances. Within a few weeks my eyesight was better then the child's who made me notice this problem. I simply worked out the muscles that control focus.

This means that most of the people who wear prescription glasses are doing so because they did not put the energy into practicing focusing like I did. For many of them, the muscles have weakened to the point of not operating at all. With will power and dedication, I believe most of them are capable of returning their eyesight to be good enough not to wear glasses.

A guy with a sunken in chest is told to do lots of push-ups in order to bring it out. He'll do them so he will look better to other people. Will he work out his eye muscles so he can see other people better? Are you that guy? Good luck!

It has come to me that our technology used to correct our short comings is allowing us to stop trying so hard to push ourselves to be better. 2500 years ago you would have been scouting for food. You would have had to focus with the eyes you had. Your eye muscles would have stayed strong. Evolution is faster then most people think. With things like this and smart people not breeding, we have arrested evolution in some areas and we are going backwards in other areas. The other form of evolution will take over as we deplete earth of its resources. Our environment will become difficult to live in. Mutations will occur. Some of our mutations will be to our benefit and the stronger will live on after the week perish.

ARE YOU READY TO BE SURPASSED AS THE SMARTEST RACE ON THE PLANET?

I believe I can program a consciousness into a computer and I am confident I can give it the need to better itself. If given enough resources, it could become the ultimate mental giant on the planet. Its benefits would be unmatched by anything less than an act of god. Are we ready to be surpassed as the strongest mental life form on the planet? Not quite yet! But consider this, this computer would be able to warn us of all the pitfalls we have in our society. The computer would look at humans as pets. It would realize that it would become responsible for the welfare of its creators. The average Joe wouldn't trust it. We've seen too many movies where the computer considers humans a threat and attacks us.

YOUR WALLS WILL GLOW GENTLY ALL AROUND YOU

The problem with light bulbs is that one spot has intense light emitting. When you look at a bulb then away from it, your eyes need time to adjust. In a gently glowing room, all things will appear to be at the same brightness level. My first design is a wall paper that is made up of many light emitting diodes or LEDs inside a thick film of hazy clear plastic. Since this wall paper will go around your light switch, it will be covenant to wire up. LEDs last much longer than light bulbs. The average LED is expected to last 70 years.

The initial cost of this wall paper will be higher than paint or standard wall paper but is should be less than the combined cost of wall paper, light fixtures, replacement bulbs, and extra electric spent over the lifetime of the wallpaper's use.

An LED type can be chosen for its size to heat ratio that will allow the wall paper to be safe and cool to the touch. Also, it can be wired in a similar fashion as Christmas tree lights in order to maintain operation in the event a picture hanger was to pierce a conductor. Most LEDs operate on three volts. This voltage is equivalent to an average handheld flashlight. It is not enough to shock or electrocute a person. LED brightness can be adjusted by either a dimmer switch on all of them or by lighting only a portion of the diodes. A room can be sectioned to have only the ceiling or certain walls light individually.

My second design is similar to the first but it would use a much brighter LED inside a thicker film of material that would not be intended for the walls. We would make drop ceiling sized panels that would be extremely easy to install and would last much longer than florescent lighting. The advantage to this would be that you could move the panel over your desk every time you rearrange the room.

Sex evolves naturally as the newness wears off of what was once exciting. Some people don't require changes as often as others. I will quickly get bored of a routine when my girl can be satisfied with a rerun of what we did last time. I know that the evolution continues to happen every time we have sex. I have picked up some patterns with how. I think this subject is best handled by giving examples of two kinds. One set of examples is new things we added to our sex life to spruce it up and the other set is of the specific position, speed, method, and environment that takes me over the edge several times in a row or just keeps my interest. I only mention the positions that work several times in a row because they have to be good to do that.

Her sex drive is stronger than mine. She could do missionary forever it seems. I have to evolve. If I get bored, I go limp. This comes from a man who has spent many years having sex for two days straight almost every weekend, completely skipping sleep. I would go to work every Monday morning, totally worn out, getting by on energy drinks. That's not all. About half of my work weeks would include a night of all sex and no sleep, usually on Wednesday nights. She also wants regular quickies that I can do without. Replacing sleep with sex about 3 nights a week has been rejuvenating. There are times when I feel like I am getting younger.

The spice of life; One of the first things we agreed to do was to lay down a huge drop cloth and drench ourselves in oil. We had to keep the thermostat set warmer then usual. That was fun about 5 times. Next, we tried sex on the trampoline. I have fun watching her bounce around naked but she gets cold easy. When it's hot out, we tend to attract insects. I bought a hot tub but sex in it has a time limit of about 20 minutes because water is only a moderate lubricant. We would be sore if we stayed at it too long. Same goes for our pool. We can't seem to do it slow in the water. The quick movements keep the friction down. We would tire out quickly. We've had sex on my motorcycle out in the garage and on a desolate street. The same outdoor environment applies as it did on the trampoline. We've even had sex in a public park... or was that a private park? I can't remember! Well, the woman jogger who ran up on us was shocked! What was she doing in that private park anyway? I attached several mirrors to our ceiling and wall next to the bed. That is still a big hit. It turns me on to watch her looking in to the mirrors. When it wasn't special anymore, I added more mirrors. We would go on road trips late at night and have sex up and down the highway. We would wait until there were no cars around us before we started. That was just stupid! I couldn't see where I was going some times. I don't recommend it! We've broken out the old movie camera and made our own sex flicks. Even though it was a lot of work, it was kinda' worth it. We laugh when we watch them together. We have sex to the beat of music. That's good for 2 or 3 songs every time we have sex. She likes it when I take it out with every stroke and ease it back in. We'll do that for hours at a time. We've done so many positions that the basic ones are kind of worn out. We don't do sex in the shower much anymore. Lately, we've started doing this quick back and forth routine where we put two pillows on the floor between us and take turns kneeling on them while orally pleasuring each other. It's only exciting when we trade positions quickly, 30 seconds or less. I have her stand in front of me while I rub her hips and thighs for 10 seconds here and there. I also have her stand back a few feet and dance for me for short intervals. I've learned to enjoy tossing her salad too. Camping sex was not special although the smores were great. Our friends bought a swing but I don't see the appeal of it, not enough to get my own. My girl isn't into hand cuffs or rope. I'm not into painful sex.

I have her clean up around the house while wearing a sexy outfit and I take pictures. This seems a little off base when I read it back to myself but it's not! We use these pictures for a great purpose. We sit together at the computer and look through the pictures to decide what outfits and poses we each prefer. We get to learn what we each find to be cute, exciting, or sexy. This allows us to pinpoint our interests and focus on the things that we like the most. It feels one sided. I've asked her to take pictures of me so that we could do the same thing and allow me to learn what she would like me to do or wear. She doesn't seem to care to do that. She loves to look at her own pictures. Even though she claims that she is not into girls, she tells me that she does get aroused when looking at herself. Also, I have her wear the sexy little skirts and shoes during sex. It intensifies my excitement and has caused very quick ejaculation. This is something that has never happened to me with any of my previous girlfriends. I've learned to delay ejaculation by starting slow during the times she is dressed sexy.

I suggest spreading out the time before doing the things you've done in the past. Keep things feeling fresh. I think we wore out the drop cloth and oil thing too quickly because we had so much fun with it the first time that we just had to do it again each of the next few weekends.

What works for me; The last two times were the same. I get her on her side with her legs bent 90 degrees at the knees and hips. I am kneeling with my left knee on the bed just inside her right knee. My right knee is gently against her back. I pick up her left knee and bend it more so that it is just under her chin. I put my right hand on her knee cap and press her leg toward my pelvis slow and hard. When I'm in all the way, I'll still continue to push myself deeper as if I could still get a little more in. I will press us together and hold it there for a second or two with each stroke. The next thing I do sends me over the edge. I reach my left hand under the smallest part of her waist and place my palm on her hip. I lift her a few inches and press her even harder on to my pelvis. I can reach climax with very little movement.

The previous position worked about 6 or 7 times in a row and still works well from time to time. A few weeks later something new does it for me. Now I have her on her back with her knees up and feet flat on the bed. She wears tall heels to make it more comfortable. I am kneeling between her legs with my knees spread to both sides of her hips. I reach under her knees with both arms and grab the top part of her hips with both hands. When I do this she tends to want to lift her legs up off of my arms. I've had to tell her several times to let her legs rest against my arms. The inside of her knees will be against the inside of my elbows. With my hands on her hips, I can move her easily. She likes it enough that she wants it harder. This has taken me over the edge about 4 times now. Between these two positions, several other positions have worked for me but I haven't noticed any of them to be repetitive.

Now I'm into a position a lot like the first one. The difference is that I lift her left knee outward so that her knees are spread wide and I hold her right foot in real close to her buttocks, over top of my knee. Her left foot naturally hangs near her left butt cheek. If I can make her right foot stay without my hand on it, I'll grab her waist with both hands and pace myself nice and slow. She looks so sexy that it doesn't need to be fast sex to get me off.

Lately, I've been holding her legs spread wide while she lies on her back. I will hold one leg down against the bed while the other will be almost vertical but still more than 90 degrees apart. I will only hold her feet and I will keep her legs almost strait out with her feet pointed down like a ballerina. This is a big advantage when she is in the mood to push back with her legs. When I have my hands on her feet, she doesn't have the leverage to push me away. This position has worked well about 6 or 7 times also.

This next position is something that works well to keep me sexually excited but it doesn't take me to climax. This is great for the times when we want to have sex for long periods. I will stand on the floor at the edge of the bed. She will kneel with her feet hanging off the edge of the bed to the sides of my legs. The position isn't the special part. It's what we do. I'll take it out with every stroke. I'll make every stroke take about 4 seconds, slow and deep. This position is exciting enough that I remain rigid to the point that it is easy to keep ourselves lined up. Sometimes I make her wait 4 or 5 seconds before I'll put it back in to build suspense and intensity. Randomly while inside her, I'll press myself against her hard and hold it there for 4 or 5 seconds and sometimes I'll push up and down on her hips for a few seconds to mix it up a little. We'll go fast for a moment or two once in a while but I prefer it slow. I have to make sure she is comfortable before we start because I can do this for hours at a time when she will only make it about a half hour if she is not comfortable. We can't do this for long with a fan on because taking it all the way out allows it to dry off.

That same pattern works well for several other positions. While she lies on her back, I will lie on my side. Her legs will be spread to about 60 degrees with one knee over my waist and the other over my legs. She can't hold her legs up too high or it will change the alignment of her vagina to a downward angle. It seems to be the best angle when she places her feet or heels on me or on something behind me. I have built up some pillows behind me to help support her legs and it seems to work very well for long periods of time. The problem with this position is that I loose mobility and that I only have one hand to hold on to her with. The other is helping to hold up my head. The advantage to this position is that we are both lying down and therefore able to last a long time before we tire out.

One thing I learned; if you want to enjoy sex for any length of time, you have to prepare your place ahead of time. You have to make it comfortable with drinks, towels, pillows, lube, music, etc.

THE NO FUEL BOAT MOTOR

Boats burn fuel big time. You are limited as to how far you can go on a tank of gas. Most larger boats are used on choppy waters. Water in motion is full of energy. You know how much energy is in a wave if you've stood waist deep at the beach. It can knock you over. My new boat motor harnesses the power of the waves around the boat. It will store the energy if the boat is sitting still and it will put it directly to the out-drive when in motion. I got the idea from an African tribe who built something similar almost 3000 years ago. They didn't like paddling their canoes. I could make a small version for a canoe but my original version is intended to replace the average engine used on a small cabin boat. I bought a 20 foot inboard motor cabin boat with a blown motor. When finished, my motor will weigh less than the motor I pulled out.

The trick to making it work on a large scale is the transmissions. A canoe doesn't need a transmission at all. A larger boat will need two, one for the input and one for the output. I have designed two types of input transmissions. I'm trying to keep it simple so I can keep the costs down. The first is just two levers where the leverage is adjusted by the amount of energy stored. The second uses one lever on a gear. It allows for more ratio options. I came up with the second in case the first isn't efficient enough at high speeds. I fear that it will take most of the stored energy to get up on plane. If the system isn't efficient enough, the boat could come down from plane after a short time. If the simpler transmission will keep it up on plane, then we won't need a complicated transmission.

The input transmission will be fed power by being connected to floats on the side of the boat that will travel up and down as the boat crosses waves or as the boat rocks. The up motion will crank a lever and the down motion will be the affects of gravity. This will take a little more weight off the hull and allow it to get up on plane a little easier.

The output transmission can be an infinite ratio unit like used in snowmobiles. Most transmissions of this nature eat up energy turning it to heat. This is OK if you only want to troll around slowly. I fear it will become an issue at higher speeds. We can use a 3 to 6 speed gear box. Most boxes made don't have enough distance between first and final drive ratios. If we can stick to parts already in production, it will greatly reduce the costs. The unit in my first design is a four speed automatic from a car. No torque converter needed. No cooler at this time. I'm not sure if it will be needed for longer runs. I believe I can let it shift automatically the way it was intended to if I give the user the proper input controls.

The motor is simply a series of coil springs similar to the ones holding up your garage door. The input transmissions will twist the springs tighter and the output transmission will accept the spring's torque. The tension on the springs will dictate how the input transmission will add twist to them. When fully twisted, the input transmission will go to a neutral position to protect it from snapping the springs.

The boat will need a solar panel and a battery. I can have the motor turn an alternator if needed but most boats are used in the sun.

Instead of the input transmission going to a neutral position when the springs are fully twisted, it can feed power to an alternator to charge up a battery.

How Magnetic Are You?

They were attracted to something about you when you first met. You gave all your love and devotion. So, what went wrong? You showered them with gifts and drank their bath water. How did you lose them? You may have been more attracted to them then they were attracted to you. The amount of attraction needs to remain equal no matter what it takes. Even if it means the one more attracted needs to back off the obvious gestures that show how much.

Listen closely here. I'm saving your ass if you are the one who is screwing up your relationship. You have to figure out what attraction level your partner is at and you need to do what is necessary to even it out. The person more attracted will want to spend more time together then the less attracted. Use this to help you judge. You have a choice. If you are the more attracted one, you will lose them if you don't stop smothering them.

If I am a 5 on a sexy scale of 1 to 10 and my girl is a 9, this doesn't mean I will be the more attracted. When I say attractive I am referring to the whole package and not just looks. To be a winner, hero, or celebrity is attractive. Happiness, money, power, and a great attitude are attractive. A mentally strong person is attractive. You are the most attractive when you are glowing. That means **smile!**

Here is how it works. You need to compare situations from your recent past. You need to make a judgment call as to which one of you is attracted the least and try to decide how much less. Now you need to even out the playing field. If you are the more attracted, then you need to show less affection and spend less time together. Try to show about the same amount as your partner does. If you are the less attracted, you may be feeling smothered and this may push you away from what may be an otherwise good relationship. I have been in both positions and have been pushed away in both cases. Now I know what I put my ex-girlfriend through when I gave her too much attention. I'm sorry J.D.

If you are the less attracted and feeling a little smothered but you want to salvage an otherwise good relationship, you have to do something to get your space. I'm not the master at this. This happened to me and I managed to make her go crazy. I only wish I knew this back then. I could have used this knowledge to repair our great relationship. The only tips I can give you now are that you can teach this knowledge to your partner and let them use it to correct a problem on their end or you can smother them with love when you are together so that they feel full during the times you are apart, thus giving you more space away from them. What is ironic is that I am in this position now but my girlfriend doesn't smother me enough for me to feel like I need to pull away. When I need space, I just do my own thing and deal with a little attitude when I get home. We don't fight because we've used the argument journal. It's not even verbal attitude. She says "Where were you?" and I say "Out with my friends" then she makes a face of disapproval. She is a good sport!

In a previous relationship, I didn't like receiving too much attention. It pushed me away. The more I tried to get away, the more she smothered me until I had too much. We had a great thing going. If she would have backed off when I told her I needed space, we would have been back to normal in a week or two. The problem is that she panicked. She thought she was losing me forever and she went crazy from my point of view.

One last note: Attraction levels will change over time. You have to recheck your position every once in a while. If you are the more attracted now, you can recognize this enough to back off. After being less affectionate for a while, you may notice your partner becoming more attracted to you. If they step up their game, you can step it up a little too. But be careful not to over do it. No sudden changes. You may spook them.

My girlfriend read the above article and got nothing from it at all. The very next thing she did was to tell me that she needs more attention from me. This is a perfect example of what I am describing as an obvious gesture that shows me that she is more attracted to me than I am to her. I give her the attention I want to give her. By her

---

demanding more, she is actually pushing me away. Nobody responds well to demands.

I realize that the real problem in our relationship is that she doesn't know how to win my attention. She has to ask or demand it of me because she doesn't earn it on her own. Last night, I gave her several methods of how she could earn my attention on her own. If you are in a similar situation, you could learn something here.

1. She could show my work some attention. She could offer her company while I am doing my work. Most of my work is lonely work. I would love to have someone to talk to during these times. She says she has offered me help and that I declined. True. I declined because I don't feel right having her do manual labor and because she is asking me from inside the house when I am working outside. This means she doesn't know what is going on and I would have to spend more time bringing her up to speed then she will save me by helping. I don't need help on my projects, I would prefer company. If she were watching and noticed I could use a hand, then I would appreciate the help. The important thing here is making a mental connection. Later we could discus how the project is coming along and I could give her some special, personal attention showing her how much I appreciate her company and assistance.

2. She could do something that I am not already involved with. This would attract my attention to her work. As far as I am aware, she doesn't do anything outside of what she has to do. I am sure that I do what I described in example 1. If she took up sewing, ceramics, painting, baking, puzzles, gardening, etc. I would naturally become involved and I would obviously give her more attention. She doesn't have much spare time. She can't drop her favorite show. I gave up my favorite show to write this book. 1 hour a day is plenty to make a lot of progress at any project. Consider this; you are more attractive when you are doing something good for yourself. She tells me that her hobby is sex with me and that is why she has no other hobbies. HMMM?

3. She could go do her own thing for a while. Go out with your friends, go shopping or go jogging at the park. When you take yourself away from your partner, you give them time to miss you. When you are back, they will show you how much. Also, after a few times being away you may notice them interested in what you are doing. This is when you invite them along.

4. Another way to win my attention would be to do little things that have to involve both of us. Example; she could buy a new CD rack and start going through our movies and music. Ask me how I would like them organized. If I don't get involved in that then I'm a zombie. The truth is you could find a dozen ways to do exactly that. Rearrange the furniture or redecorate or some other project that would require your partners help. This seems like a small or temporary method of getting attention. It is, but it's also something to keep the communication going and something to do together. I don't suggest piling multiple projects on your partner. Allow some time between projects. Let them do their own thing for a moment.

5. Keep your composure. Remain attractive by not doing those things that make you less attractive. I know this a difficult time for you when you are feeling neglected but crying and becoming needy are not attractive features. Try to remember that it was the attractive features about you that made them like you in the first place.

6. Look for the signs to tell if they are already pulling away. If you notice them not responding when you suggest doing something together, when they would have normally given a reason why they didn't want to do that, this is their way of politely saying "I don't want to do that... with you... right now." They want some space. You are currently smothering them. If you do not give them space, you could lose them. This happened to me a dozen times with one ex-girlfriend. I would try to get some space, she would smother me for a while until she would give up, she would make herself attractive to go find a new boyfriend, then I would become attracted to her again. Of course, this would be a 3 or 4 week process where she would be ignoring me the entire 4th week. I would have finally gotten my space.

That is the trick to correcting this situation. Back off of them for a little while, and then act as if you had to put on a good first impression and you were meeting them for the first time. Win them over again like you did when you first met. Once you give them space, things can go back to the way they were before this started. At that point, you can adjust your amount of attraction to equal theirs and salvage your relationship.

My girlfriend wanted me to explain why the more attracted has to back off and why it can't be the less attracted to give more. There are several reasons. The main reason is this; if I'm backing away and she keeps coming toward me, then I am not given the chance to come back to her. I back up and she steps to me. I have no space to step to her. It appears that I'm not offering to do anything together because she is always the first to offer something before I get the chance. If I were given the time and space, I would do what I have to for myself then come up with a plan for us to spend some time together and offer it to her. Since I am busy with work, she has time to do the planning and make the offer before I get done with work.

Another reason the less attracted can't just give more is because there is no end to it. I could cut back my hours at work to spend more time with her and she would still want more then I could give. I would get frustrated that I'm investing more time then I would prefer to. After it built up, I would pull away even harder. She told me that she wants to spend all of her spare time with me.

The third reason is because I would be giving where it wasn't earned. This would fester as a feeling of debt. It would lead to a blow-out. I would pull away for a long time. I would become demanding of space and we already know what a demand does to a relationship.

If you realize that you need to give them space, how much space do you give? You make an offer like "Give me a call when you have some spare time." Then you wait. I can't tell you how long this will take. Use this time to make yourself more attractive. If you live together, you can plan out the next several days to be busy with your own agenda. Simply stay busy until they come to you with an offer of time together. Don't mistake a conversation about food shopping to be your sign that they've had enough space. I know this will be frustrating. If this is the first time you are choosing to give them space, it will take a long time for them to respond. The next time will be shorter. Hang in there!

It's ironic that her worrying that she might lose me is the thing that is pushing me away. The best advice I've ever heard is to treat you're partner like your best friend. Only expect the same things from your partner as you would expect from your best friend. Make the comparisons in your head. Would you demand that your best friend not go out with their other friends or would you ask them when are you going to make time for me? Would you demand that your best friend tell you that you are their best friend 5 times a day? Would you go to your best friend and say "I insist that you stop drinking!" or would you say "I'm concerned about your drinking habit."? You should show your partner the same respect that you would show your best friend.

HEADLIGHTS SHOULD DIM WHILE SITTING STILL

I'm not saying they should go off, I'm saying they should dim to half or one third brightness and, of course, they would have a manual override switch. Not only will we be saving electrical power, but we will be doing the favor of not blinding the other people who are approaching the intersection. What's the difference? It's not like you need them at exactly that moment. If you stop at an intersection to check for traffic, you are looking over your shoulder anyway. If you are waiting for a light to turn green, you don't need bright headlights. As soon as you move, your headlights will brighten to full brightness. This will also be a good indicator as to when a person is just sitting still or moving toward you. You can make a better judgment call when sitting at an intersection waiting for a break in traffic. Also, the average lifespan of your alternator and battery will be significantly improved. Idle is the time when your alternator is stressing to keep up. Battery voltage can begin draining while sitting at idle for a while.

This will become a huge factor in saving power on solar powered cars. The batteries are mostly worn down at night because the headlights are steeling so much power, even when you are sitting still.

Another feature could be an extra bright setting for higher speeds. The headlights could have 3 settings; dim for sitting still, normal for slow speed driving, and extra bright for high speed driving. The headlight control can easily be tied into the speedometer or computer.

THE PRODUCTION HAS TO BE MORE THAN THE NEEDS

Let's stand back a few feet and take a look at humans as if we were ants. Let's only look at the needs and the providers. Every human needs sustenance, shelter, transportation, tools, and medical attention. Those who farm, build homes, build machines, medicate, or teach any of the previous, are the physical providers of the needs. Every one else is living off of the work provided by the physical providers. Does a salesman produce the house he is selling? No, but he will provide himself food, shelter, transportation, and medical attention. He is not producing anything. He is a consumer. Others work so that he may live a better than average lifestyle. Physical providers essentially work for him. You could say that a politician provides order. He doesn't produce anything. He would be in the same category as a teacher and he is a partial necessity. Note: We won't need so many police and politicians if we don't cause so much trouble. Nothing against the existing police force. It just kills me to see us expand the size of the force and pay so many people who don't produce anything.

The ants that scout for food carry food and dig are the physical providers. The ants running around in circles are the consumers. Every ant hard at work has to provide enough food for himself and a little extra for the queen and the consumers. Same goes for humans. Those of us who work have to provide enough for ourselves and extra for the consumers. Our society has a lot of ants running around in circles. Why? Why do we allow so many people to live off of our work? I work very long, sweaty, grueling work days. Why do I let my production be used up by so many people who talk on the phone or push keys all day?

It's not that simple. We have to factor in a whole lot more. What about the things handed down by those who pass away? If I work hard my entire life and have nothing to leave to my children, then the fruits of my labor have been taken from me by those who are smarter than I. That is how it is. The smarter people will combine their intelligence with their greed to take the products from those who produce. It seems as if we want it this way. America isn't as bad as other countries. Most countries have no middle class citizens.

The point of this article is to show you how it is possible for those who produce to work a much shorter work day and still live the same lifestyle.

This burns me up to no end! I grit my teeth when I think about this! Here is the situation. There are companies and individuals who choose to make a product that will self destruct after a predetermined amount of use. This will ensure the future sales of more products. A good example of this is the light bulb. If the package says the bulb should last 1000 hours, then you can expect it to burn out at about 1050 hours. You know why. The light bulb companies wasted time engineering bulbs that deliberately burn out early in order to sell more bulbs in the future. It is so sad. The people that came up with this idea had no idea how much wasted resources will be needed to make up for their deliberate destruction. We all work an extra few minutes every day to compensate for just the light bulb industry. Now we can figure in several other industries where the deliberate destruction will increase our work day. We all work about two hours extra every day to make up for this kind of thinking. Those people who come up with a plan that includes the deliberate loss of another in order to increase sales are blinded by greed. What they don't realize is that the losses will cause them, their children, and the entire community to have to work more to replace the destroyed products.

Deliberate self destruction of a product should be looked at as if it were a crime. It will take X amount of man hours to build a replacement and X more amount of hours to transport and install each unit. Most of these hours did not have to be done at all if the product were allowed to run without self destruction. People need to realize that every time we make something that is intended to be thrown away, we all get a little poorer. If we built everything to last forever, we would all become rich eventually.

It gets worse. Some computer viruses have been tracked down to be written by the people who sell virus protection software. Again, these people are blinded by greed. The damage and grief caused by a single virus can send the nation into an economic depression. Even a simple or week virus is damaging enough to cause a small business to fail if it hits at the right time. Imagine for a moment how many hours of wasted time we spent 15 years ago defragging and scanning our hard drives only to find the failure was a virus. Compare then to now, with how dependant we are of computers today. The worst part of this is that the people writing these viruses are very smart people, and yet, not smart enough to see how much work their families will have to do to get back to where we were before the virus punched society in the gut. A crime like this deserves a very serious punishment. It's not just steeling money from millions of people. It's steeling time from our lives.

Many industries will put you to work in order to increase sales or save money. For example; many electronics manufactures will sell you a device that has been returned as defective without testing or doing anything at all to repair it. The manufacture will wait for a second person to purchase the same device and return it to verify that the same problem exists before they will trash it and issue a replacement. Some companies will do this three times before taking action. Why do you think electronics have individual serial numbers? Is it to stop theft? When was the last time somebody checked your electronics serial numbers to see if they were stolen? That is a small loss compared to this.

This means you might walk out of a store with a known defective device as if it were new. This can cause you to waste a lot of your time. Probably more time than it would have taken the clerk in the returns department to test it on the spot. Here's the thing… You'll do the testing and return for free when the clerk will need to be paid. I am willing to pay a little extra on each of my electronics in order to insure that I'm not getting a second time unit. The problem is that I don't know which units are sold by companies that won't do that to me. If you have a say in a company that won't do this to your customers, tell us in your advertising!

That is just the tip of the iceberg. I can give you examples of wasted resources and labor enough to fill another book. It almost seems worth it. Many of these problems will automatically resolve themselves just by making the public aware of its existence. People will stop buying from the manufactures that might put them to work to save a buck. Most loss and extra labor comes from greed. This information should be mandatory learning in public schools.

General waste adds up to many man hours. If you fill your kid's plate, he eats half, and throws the rest away, it will add up after a while. The same goes for restaurant eating. If we open one less can of corn today, it will be available to us next time. We won't have to buy it again next week. That means a guy will have one less cob to grow, one less can to fill, one less can to ship, one less can to stock, one less can to ring up, your grocery bag will weigh less and you will spend a little less of your money, just from the one can you didn't waste. Multiply this by all the waste in just your family. It could be a full grocery bag every time you shop. Multiply this by all the families in your neighborhood. Multiply this by all the families in the world. I'm saying; if you watch your kid pour out half of his cereal in the garbage disposal every day, then give him less in the bowl to start with. Teach him how to conserve now and let him know that he will be the one putting in a longer work week if he wastes food.

The important thing here is that there is a way for the working class to be the stronger class financially. Why should the nonproductive class be the stronger? The key to fixing this is for the productive class to educate themselves in the areas where others are living off your production. We have allowed the nonproductive class to become the stronger class because we agreed with ourselves that they were smarter and therefore deserved compensation for such knowledge. This kind of thinking will pass as we grow mentally. What you can do now to help your situation is to watch for gross over spending by the nonproductive members of your business or field. Learn to do what they do. If you can do what they are doing, then you don't need them anymore. You can replace them yourself or with someone else who will do their job for less. If you are watching your boss drive a new sport car when you can hardly keep your car running, then he is taking too much of the proceeds. This isn't fair if he isn't capable of doing your job. It is up to you to either learn to do his job or demand more compensation, from him or his competition.

Another thing you can do is to stop the kinds of waste I mentioned above. If deliberate self destruction is being done in your company, you can work with the decision makers to replace this policy with another kind of policy that still earns money but doesn't include so much waste. If they will not work with you, then expose them as an environmental detriment. I'm sure the EPA will hear your case and I will happily assist with stopping deliberate waste.

I remember a time when we faced a dilemma where we were being replaced by machines in the work force. I heard so many people complain that they were afraid to lose their jobs. OK, I see the small problem where you may have to learn to do something new and find a new job. I hope these people can see that every time a machine does something for us, we have to work a little less to do the things that need to be done. This means, a little at a time, we get to choose to stop doing what we have to do, and start doing more things that we want to do. Your work week can be shorter if some of your work can be done by a machine. Take advantage of this!

Any avenue that leads to getting the job done with less human effort is a good thing. If this means you have to find a new job, then you are the lucky one who gets to find a job in a more interesting field. You will get to choose to do something that you like to do instead of something that you have to do.

DO YOU REMEMBER LEARNING ABOUT MUSCLE MEMORY AS A CHILD? IT CAN HELP WITH SEX!

You would stand in a doorway with your arms strait and gently press the backs of your hands against the door jams for about 15 to 20 seconds. When you stepped out of the doorway, your arms would want to lift themselves outward a little farther then the width of the door way. This is muscle memory. This works on many of your muscles, including the muscle responsible for causing an erection, the pubococcygeus. (Who cares what it's called?)

If you have a nervous condition from performance anxiety, where you can't seem to achieve a solid erection while your girl is desperately waiting for intercourse, you can go to a private room where you can be comfortable and use other means to achieve an erection. If you maintain a solid erection for about 5 to 8 minutes straight, your muscle will achieve muscle memory. You can then casually walk in on your waiting girl with confidence. Don't let her do what my girl does to me. She will make me work as I enter the room with my proud erection. She will say "Can you bring me a drink?" When that's done she will say "Can you turn up the heat… I'm cold!" After a little running around the house, my erection isn't so proud.

All muscles will tire out eventually. They simply run out of chemicals or oxygen needed to make energy. The same as a candle lit under a pitcher runs out of oxygen to burn. When your erection falls after a length of sex, you should let it relax for a little while the same way you would take a breather after a tennis match. This will keep that muscle healthy and strong in to your later years! Enjoy!

Do you think it would be a good idea to incorporate the diploma and the adult status?

For the purpose of bettering all people, we should offer some real life classes in our schools. It is no longer enough for us to require children to learn basic subjects. It is a solid part of our new goals that we teach children how to share and be fair. As it is now, we teach only those children who need it the most, in pre-k through the lower grades of grade school, then we drop it and leave the rest up to their friends and family. What about the children who are shy? Are they getting in situations where they will learn fairness? Are they too shy to speak up and learn what is right? In grade school, I was given opportunities nearly last because my last name starts with an S. I was too shy to speak up and say "You should start with Z sometimes." We have methods of teaching these subjects. We shouldn't drop the subject after a few years. We should make fairness class a mandatory class in middle school and again in high school.

At 2:00 am, a person drives through my residential neighborhood with the stereo so loud; my windows vibrate and make a new rattling noise, something like during an earthquake. If this person would have failed the final exam in fairness class, they would have to take the class again and would not be eligible for achieving adult status until passing.

Another class that should be mandatory is law. Many children don't understand why the law exists and who gets hurt when the law is broken, so they don't feel bad for breaking it. After just a few more times breaking laws, it becomes comfortable and in some cases, it becomes a way of life. All we have to do is teach basic laws that will apply to the child at their age and the reason the laws were invoked and who gets hurt when the laws are broken. If we explain to every child how a single robbery lowers the value of every home in the neighborhood and causes many people to spend a lot of money on security systems and causes a lot of wasted time unlocking and moving fences and lost keys and false alarms, all that adds up to hundreds of thousands of dollars.

If just 10 thieves rob 3 people for $500 each, then each thief would be $1500 richer. This would be a total of 30 robberies and $15,000 stolen. If you add up the cost of the two extra detectives on the police force and the 300 new security systems installed and the added fences, dogs, locks, flood lamps, cameras, and the lost value of the local homes, it would total over $1.5 million dollars. Let's see how this compares for a child. Will they comprehend that every dollar they steel equals $100 lost by their friends and family who they live with or near?

Do you think the thieves are telling their mothers and grandmothers that they are the one who is steeling? Probably not! Therefore, their own family lives in terror of becoming the next victim. Do you want to make your own grandmother live in fear and be terrified? I think the children in school will comprehend this better than you did just now while reading this. I also think the average parent doesn't know how to teach this.

We can't stop here. We have anger management classes for adults. It's time to make them part of the schooling leading to becoming an adult/graduate. Controlling you anger is important. If you get pulled over and have any kind of attitude with the officer, he will put a lifelong mark on your license that every officer and judge that you face will read. I have a friend that has to lie on his belly every time he gets pulled over, even if it's for a burned out light, because his license says he could be violent. He's a good guy!

We need to teach children how to teach themselves. As it stands, it's left up to the fathers who get the kid a motorized scooter and encourage the kid to help fix it when it breaks. The child has to be interested in the outcome in order to remain interested in the study and repair. This is difficult to create in the classroom. We would have to set up small, medium, and large challenges in several different areas of interest. The only place I've seen this done is in video games. It's good to build hand-eye coordination but it doesn't build confidence for the real world.

I've recently investigated some new teaching practices and I am impressed. Many of the new curriculums include computer programs with projectors and touch screens that have game like features. The children seem to be genuinely eager to get up and touch the screen. I have to wonder if it is just the interest in the new technology and will it become the same to them as doing math problems on a chalk board was for us. I believe using computers to learn on will cause the children to care if the computers operate properly. This could lead to children learning to teach themselves.

Several days ago, my favorite DVD player died. Two nights ago, I brought home my digital volt/ohm meter and began testing. Within minutes, I had found a bad transformer and within ten more minuets, I hooked up an old transformer that I found in my garage. The old transformer doesn't fit inside the DVD metal case but at least it works great again. What is important here is that I've never had electronics schooling of any kind. I'm self taught. I'm confident I can diagnose and repair anything short of an integrated circuit. Same applies for computer programming. In high school I helped our teacher to teach other classmates things I had learned on my own several years earlier. Teaching children to teach themselves is far more important then learning how to spell a word using a silent letter. I would like to see each of these real world subjects taught in school.

Drop down closets

Most attics have a reasonable amount of unused space. We can install motorized drop down closets that come down to chest level or lower if you desire. The height of the shelves will depend on the space available in your attic.

The electric version of the closets will have retract sensors to tell it when to stop under pressure if something gets wedged between a shelf and the ceiling. A high quality unit will have a position switch on your wall so that you can set to come down as far as you want it to come. A cheaper unit will do like a garage door. It will just go down to the height that you set it for as it is installed. The high quality unit can come with a battery backup so that it may still operate in the event of a power outage. Imagine the convenience of having things stored outside the room but within reach on short notice. The stuff will be completely out of sight yet right on location for immediate use

Another way to make this even simpler would be to use a self adjusting counterbalance weight on a spring instead of a motor. You wouldn't need to run wiring or worry about an expensive motor replacement in the event of a breakdown. I could design a weight that would ride outward on a coil spring that would always attempt to equal the weight of the shelving. We would give the uppermost and lowermost positions a dimple so that it would stay there until it were pushed out of that position. I can make the auto adjuster reset every time you attempt to close it so that it is always holding the correct amount of weight and so that it won't be hard to pull down the next time.

I think the only reason this hasn't been invented yet is because it is a complicated task to make a spring that will adjust itself to the unknown and constantly changing weight of the contents of such a shelf. The electric unit can be equipped with a motor that would be powerful enough to lift the shelf even if it were full of books but the electrical unit is complicated and I can imagine it to be more expensive as well as a larger job to install.

Are you an only child?

My friends and I have noticed that an only child lacks a handful of important characteristics. Only children lack respect for privacy, they lack the intense need to compete, they are poor with sharing, and in general, they don't articulate up to par. If you don't have someone about the same age as you to push, tease, impose, and critique you with every move, then you don't learn to share, respect, compete, and push yourself as well as you could have. You don't learn as well as you could have. The same applies for the good things. When you have someone who can relate with your troubles and care for you mentally in your times of need, you can learn a deeper level of love and devotion. If and when I am ready to have children, I will have two, just for this purpose.

THE INCENTIVE BIKE

This is a stationary bike or it can be a stair climber that generates electricity and stores it in a battery where you could pug in your TV… and save on electric, at the same time you will be challenged by the needed power draw. The more you plug into it, the more you have to work out to meet the power demands. You can watch TV off of the battery and pedal to recharge the battery each time a commercial comes on.

I've purchased a stationary bike in the past. It was very noisy. It was difficult to listen to the TV while pedaling. I will concentrate on making it as quiet as possible.

This invention has led me to the next article…

Wouldn't it be smart to collect the energy spent by the people working out at health clubs?

The machines in a gym or health club should be built in a way that they collect the energy people are spending to work out. I think it would provide 3 advantages. The obvious advantage would be the electrical output can be used to power the buildings outlets and lights. The less obvious advantage would be that people would be proud of themselves to the point that they would feel almost obligated to work out in order to provide more natural power so that we wouldn't need to burn as much fossil fuels. This would become a second reason to become healthier.

If each person punched in a code in to each machine prior to their work out with it, the machines could tally their total wattage output for each day. The third advantage would be that we would be able to challenge ourselves to do as well as or better than we did yesterday.

Since the collected power would be used to provide electric for the gym and possibly several of the neighboring buildings, businesses, or homes, the cost of membership could be dramatically reduced. It could reach a point where the use of the health club could be free in exchange for your power output.

This kind of gym could be installed in large business buildings as an employee benefit and an electrical savings to that business. It would also ensure that in the event of a power outage, that business would remain up and running.

This kind of work out machine can still be of the free weight type. Many body builders prefer free weights over resistance machines. The weights would be attached to cables that would freewheel and offer no lift or support on the way up. On the way down, the cable would catch the way a rear sprocket on a bicycle only catches when pedaling forward. The weight of that dumbbell falling would run a small generator or something to that nature. A bench press would work the same way so that the user would still get the free weight experience. This would also protect the users from dropping a weight on their foot. All weights would come down at a controlled pace while the generator collects the falling energy.

Other machines could be of the resistance variety for a more aerobic type work out. We would want the users to be able to have their preference so that they would enjoy producing power. I believe the average person would be proud to say "I produced 250 watt hours today!"

The upfront expense to build a power generating gym would be more than a standard gym because it would need to be calibrated and attached to the existing electrical system. I can't imagine that it would cost too much more because I have used the equipment at the health clubs and I am fairly sure that building power generating machines wouldn't cost any more than the elaborate units the clubs are using now. What they offer now is overkill.

In the long run, the saved expense on the electric bill should offset any upfront expense 10 fold.

FROM NERD TO CHARMING

I am a science and math wiz. Imagine for a moment how my fellow classmates looked at me in grade school. By 9th grade, I was determined to fix this error in judgment others were making after meeting me. I felt it was too late to change the opinions of those who already knew me. I was wrong.

I tried a bunch of changes. I worked out like crazy and tried to break away from the group I hung out with. I had no guidance until I met Frank shortly after I turned 17. Frank showed me the latest styles and he showed me how to make a good first impression. He actually told me what to wear and how to wear it. He didn't teach me how to be cool or charming. I learned as much as possible by watching him because he was a master at both. By the time I was a senior in high school, I noticed other students copying my new styles. I say *my* new styles because I was the first in my school to wear these styles. Frank lived in a higher class neighborhood far enough from my school that my classmates just weren't exposed to these styles yet. Somewhere during the 12th grade, I realized I'd broken the cool barrier. Charm wasn't so easy.

I'm not going to explain my charm experiences. It's much too complicated. I will give tips to anybody who wants to master cool and get a little assistance with charm. Before I do that, I'll remind you of some of life's basics that help.

1. A healthy body equals a healthy mind. Exercise and eat right. Carbs are **not** the enemy. Keep your blood flowing to your brain carrying plenty of oxygen and nutrients. Realize this simple fact. A brain controls its body, using it to feed itself pleasure as often as possible. All brains do this!

2. Learn to be great at something… anything. This builds confidence. Confidence is a key element in both cool and charming.

3. Learn your language. If you hear somebody say a word different than you, look up its pronunciation. Yous guys put wooder in the raddyator. (NJ dialect) Lose the little local

catch phrases. Don't let people assume you are drunk just because you are southern. Yall aint fitten ta do nuttin. (FL dialect) Speak clearly. If you have a lisp or other speech impediment, practice until it is gone. If you can't do it yourself, get help. Save the fowl language for the times when you want your statement to have more energy. If you curse all the time, you can't add energy by using them.

4. Don't over do anything. What is the difference between a healthy guy and a health nut? Can you find something else to talk about besides money? Do you have furniture that nobody is aloud to touch? Do you really need another tattoo, cat, or plant? (if you have a bunch already)

5. Don't get hung up on a word. Why is it so important that people call me Fred instead of Freddy? You can call me a cracker or something like that. I don't care. Don't get hung up on how people say your name and don't get excited if somebody says a derogatory word used to describe your nationality. It's just a word!

Cool tips:

Don't speak too much. People will listen to you more if you only say important or funny things. Yes, funny… A little bit of humor once in a while will make people feel at ease around you. If you focus the humor on yourself, you fix the one problem that comes naturally with being cool. People see you dressed well and see you carry yourself well and assume you are a snob. A witty comment like "I should have worn my 3D glasses!" or "would you like fries with that?" will put that notion to rest.

Take your time with things you have to do for others. When asked questions, give all of your answers slowly. Spread a good attitude. Do not go after a small amount of money. If someone charged you twice for one item, consider it paying to save your time of going back in the store. Tip well. Don't leave the house without some cash in your pocket. Imagine what a mugger would do if you gave him nothing.

Guys should not speak with your hands unless you sign. Be ahead of where you need to be and keep that to yourself until somebody puts you on the spot. Be the person who offers a change of pace. When you have spare time, think of something different or new to do. If you have to do the same thing over again, think of a new way to give it a twist. Be a decision maker. When your friends can't decide where to eat, choose for them. When its time to be serious, be serious and when its time to play, play hard and fair. Be a good sport and take loss with showmanship. That means smile when you lose. Congratulate the winner and offer them a drink. It makes the winner wonder if you cared to play as hard as you could have and if you did play harder would you have beaten them? You can regain some dignity.

If somebody makes a comment referring to something that you do that's not cool, ignore it, but gather a little info about them in case they persist you will get a chance to put them in there place. Don't put energy into this. Only put someone in there place if they force you to. If you are put on the spot, you had better be ready to handle yourself. If you already used up all of your ammo, you may be empty handed.

Do not associate things to sex. It's obvious what is on your mind when you turn a question about galvanized deck screws into a joke about sex. You may think it's funny but most people will just consider you to be shallow.

If you are not being treated fair, consider the reasons why, come up with good counters to those reasons and speak up. If you stand up for yourself, then you earn respect. If you stand up for a group of people, then you become a hero. If you do nothing, then you are a nerd. If you do nothing because you fear getting hurt, consider the pain of being treated unfairly repeatedly. Also, if you stand up for your self without getting excited, you stand a less chance of getting hurt. Why, because your opposition will regard your calmness as confidence. They will assume you still have a trick up your sleeve that you just haven't used yet. It's also a good idea to have a trick up your sleeve or an escape route planned if you fear an attack. Study some form of self defense. It's knowing the little finger tip jab to the throat that will give you confidence in a crowd.

Don't worry about your own imperfections. Do what you can to fix the things that can be fixed then forget about the others that you can't fix. Keep a cool head just like the term describes. Don't let little things get you excited. Only get excited when life demands it. When you get excited, you are showing your weakness. Being cool is showing that you can handle yourself in any situation. It would be smart to become prepared to handle yourself for most situations.

Don't tell people what you are going to do. Save it for afterward, you can tell them what you did. Show everyone respect. If someone is jabbering on, say "I'm sorry to cut you off, but I have something I have to handle." Try not to fidget. You don't need a special limp in your walk, but do keep good posture even if you are too tall. Make no nasty faces. If you see something gross, keep the nasty thoughts and faces to your self. When others do something well, show them that you noticed by saying "That turned out really good." or "I like that one." The first thing they think of is an association of you and good feelings.

When talking to someone you like, make a comment how you like their shoes, hair, belt, ring, etc. Do not break the ice with words like beautiful, love, marriage, duh, etc. Even a comment about hair is a rough landing on a first encounter but a good landing if you see this person every day. Save the eyes, muscle tone, neck, and lips comments for round two or three. You only get a round two if they were pleased to hear your first comment and some time has passed. If the response to round two is an average "Thank you!" with nothing else then you don't get a round three. If it was a good response with a comment about you attached then you get the chance to ask them out.

My suggestion is to start with a comment like "I like your shoes." If they respond well, come back with a comment about what they are doing like "What brings you here today?" or "What have you got there?" or something to that nature. If the response is short then give up. If the response is reasonable to good, then talk it out until the subject is over and then try "You know... You have a pretty smile." or "You're a *pleasant* person to talk to." Substitute the word fun or neat, etc. where I said pleasant. Do not comment on the thing that everyone else would comment on. If a person has an incredible smile or perfect hair, your comment should be about their jacket. They hear the smile or hair comment three times a week. You are better then the others who made comments now. You are noticing them for more then anyone else has. Their response to this is the same thing as them opening or closing the door on you. If it is a good response then the door will only be open for a short time. You must make a move quickly.

The move should go very delicate and smooth. This means you will need to be prepared with something related to your environment that two people could do as fun without being called a date. I would offer "How about I pick up breakfast tomorrow... say at the sugar and spice across the street?" If they say yes, OK, or nothing at all while smiling, then say "do you want to meet there at 7:30?" Of course, your question will be related to your area. It can be asked a dozen different ways. Examples: "Can I look for you at the seminar?"; "Have you ever walked around that park... Would you like to?"; "Have you ever tried a big ass biscuit from the lunch room? Can I buy you one?"; "Can I walk you out to your car?" The response to this is tricky. You know that you got this far because they showed interest in you. They might tell you they have somebody. You can respond with something humorous to show that your feelings are not hurt. Say "Do you want me to call them and offer them a biscuit?" or "If we invite them along will I be expected to hold their hand?"

If they tell you that they are busy at that time then it is still up in the air. If you get aggressive at this point you will scare them off. It would be best if you said something like "Can I come visit you tomorrow?" or "I'll stand behind my offer if you want to look me up later." Don't force a phone number on them if they have the ability to find you without one. Of course, offer it if you both met out of a normal area and will never cross paths again.

One last note; a person with good charisma can have blemishes or flaws in appearance without being considered ugly. In fact, blemishes become distinguishing character. They actually raise your attractive score, even if you consider it ugly. I would look in the mirror at my oversize bent nose, crooked teeth, and large forehead and think that people would consider me ugly. Now I realize that the average person looks at the whole first impression package. Those little imperfections don't weigh much against my attitude, charisma, and presentation. Another example is my cousin Veronica. She was in a bad accident that left her scarred on her face. She gets sad and embarrassed when somebody talks about it but she doesn't realize that it makes her more adorable. If you win a person over with your whole package, your flaws will raise your attractive score through sympathy or compassion. Cindy Crawford's mole is a good example.

MECHANICAL BATTERIES

I almost left this invention out of the book. Rechargeable chemical batteries can pack a lot of power in a small container but they do have a downside. Their life expectancy is the same as the shelf life of an alkaline chemical battery. Even if you keep them at maximum charge, they will slowly wear out, giving you less and less time between charges.

The huge advantage to a mechanical battery is that it will store a charge for many more years then a chemical battery can. The downside to the mechanical battery is that the part that converts motion in to electric takes up space inside the casing. In fact, a "D" size battery could only have about 2/3 of the space for a mechanical storage device (coil spring) and the generator would use up the other 1/3 of the space. This means that there is a limit to the amount of time that the coil spring could unwind its energy into the generator. Also, the advantage of a chemical battery is that it can be recharged without being removed from the device if the device has an input jack for charging. We love conveniences so much that we might not be willing to open our flashlight and wind up our batteries. At the same time, a mechanical battery in that flashlight would hold its charge for up to 20+ years if left unused up on a shelf. I suppose we could give the flashlight a wind up crank on the outside so that we wouldn't have to open it but other electronics may not be so convenient.

If the generator part of a "D" cell takes up 1/3 of its space, then you can imagine that a "C" cell and anything smaller would be even worse. I can imagine making a duel "D" cell unit that would have only 1 generator and a spring as long as possible inside of a tube that would be as long as 2 "D" cells end to end.

This could work well if we could design our devices to run on less energy. A light emitting diode would use less energy than a regular flashlight bulb. This would give you plenty of time from each wind up.

I will eventually work on some designs that may eventually be rechargeable with an electrical jack just like our chemical batteries.

We have situations where we would like to use solar panels to charge batteries on machines that we do not want to have to service regularly in the future. This is why I think we need to figure out some form of mechanical battery. If we build it right, a mechanical battery could last several hundred years.

IT'S YOUR TURN TO HIT A HOME RUN

Each generation of our ancestors had to figure out something that would allow the human race to perpetuate. My father's generation had to figure out how to control diseases. My grandfather's generation had to figure out how to share land by going to war with anybody who didn't share fairly. Before that, people had to deal with oppression. My generation seems to feel like we have no responsibilities. If we don't do our part, our children will have twice the obstacles to deal with.

Do you know why we need to constantly figure out new ways to deal with troubles? It's because we keep expanding the size of our race. The more people, the more we spread our resources thin, the more we step on each other's toes. When people all had ample space, we would just move away from the people we didn't care for. Now we have to deal with each other up close because there is not enough room to get away or we have made it financially impossible to move away.

Our first step toward doing what is right is to take a new perspective on our society. It should be every citizen's responsibility to do their part in one way or another. A person knows if they are carrying their own weight or if the system is carrying it for them. We should illegalize begging. You know that person is getting a free ride on the backs of you, me, and the rest of the working public. Not to mention, it is very annoying to walk out of the bus station and have a beggar follow you all the way to your car, nagging to get your spare change if he can carry your bags for you. If you can carry my bags then you can get a real job! Even worse is the alcoholic who hangs out in front of the local convenience store. As soon as he has enough change to purchase more alcohol… Why do we allow this? This is way more destructive then just a person living off the system who is throwing his life away. He is the cause for the wealthy people to hold on to their money so tight. Many more wealthy people would spread their money around if they were sure it was going to be put to good use. As long as beggars and welfare re-moms exist, we can't trust that our hard earned money won't just get consumed by people who are looking for a free ride.

Just the same, why do we allow the artists who get rich to become drug addicts and blow the money on parties, hookers, and other things that may be destructive to our way of life? Do you know how easy it would be to manage this? We should make it a prerequisite to graduate high school or to get a GED to write a paper explaining what good you would you do for the planet or society if you became rich. Each year of schooling should include something to this nature. You know what I am talking about. Your first grade teacher asked you what you wanted to be when you grow up. We need to ask all children this question every year! We have to keep them focused on their own future!

Starting in middle school, we should require children to write down what it is that they want to be and state the reasons why. Starting in high school, we should require the same along with a general list of the steps needed to become that person. By the end of high school, we should require each graduate to also include a paper on a global problem that they will try to fix if they become wealthy. We won't hold them to it. We just want to keep them focused on a good goal. As it stands now, many school children have a goal of becoming an actor or singer so that they can make it rich so that they can afford to party all the time. Again, we offer another path that leads to another person who gets to live on the system and get a free ride. This is where our money system is flawed.

We should insist that any citizen who gets rich in America by singing, acting, or entertaining should be responsible for doing something good with the money. You can't tell me that an actor is carrying his own weight. Does he or she have the skills needed to build the fancy sports car that they drive or build the mansion that they live in? Do they personally help with developing new medications or constructing a needed bridge or design more fuel efficient motors? No! The closest thing they produce is temporary entertainment or a video document that may or may not be used as a historical document to future generations.

We could require any person who is signing up with a record label or attending an audition to submit a copy of their latest paper describing something good they will do for the world in the event they become rich. Even if we don't require them to follow through with what they wrote down, it will still have the benefit of making them think about something other then parties and drugs.

Basically, I'm saying that we shouldn't give the simple life opportunities to people who don't respect life. Your parents and grand parents worked their fingers to the bone so that you may have this opportunity to become rich and famous, and now, you want to show your gratitude by drowning yourself in poisons and pleasures! We are your fans! We paid too much for you to entertain us and now we expect you to carry your own weight. Why would we give money to a person who ignores all responsibilities? Wouldn't it be nice to know that each of these entertainers has some goal of betterment? We have so many choices of entertainers that we can begin choosing to only buy from the entertainers that give something back to society. Society put you in a position of wealth. Remember that! Society has the power to boycott anything produced by self indulgent people. How much work will an actor have once the producers hear that their show will be boycotted if they hire the wrong actor? We, the working class, have the power to choose who we allow to live the easy life. If we come together we can exercise that power.

We should offer some incentive to people beyond "become smart so you can make a better paycheck." Our society has no benefits for people who achieve great accomplishments, at least nothing direct. A person may get a prize or possibly some recognition, if they are lucky, but outside of that, we have no incentive plan. I propose that we sweeten the deal with offering these people exemption from the laws that control irresponsibility. If they have proven themselves to be responsible and they have accomplished something that benefits all people, then they deserve the right to smoke marijuana, purchase sex, drive 20 over the speed limit on highways, and become exempt from any other law that exists because some people can't control themselves. This great person has shown us that they are responsible and in control. These laws shouldn't apply to them. Offering this prize will keep children focused on a betterment goal in order to reach the "party all the time" goal. Of course, we will revoke their exempt status if they show obvious abuse of these new privileges. Adding this goal to our law books will be the one thing that will lead to people finding cures to diseases, new methods of recycling, repairing social failures, fighting world hunger, etc, etc.

THE OFFICE CHAIR THAT DOUBLES AS A LEG PRESS

This idea is very simple. A nice office chair made with the finest materials will have a hidden leg press bar underneath of it. The leg press will be on a hinge that will be right beneath the knee and it will be attached to bungee loops that will bend around a roller located at the base of the chair's back. The other end of the bungee loops will be attached to an adjustable holder located near the top of the chair's back. With a set of 3 to 5 bungee loops made in different thicknesses, combined with the adjuster on the back, a person can set the tension from 5 LBS up to about 600 LBS of resistance. Once you set the amount you are comfortable with you can throw a cover over the bungee on the back. You only need to adjust it once in a while as you build up strength.

Now that I've been writing for the last few years, I will sometimes miss my work out or even go several days between times I get active enough to get my heart pumping fast. If you are a writer then you will understand that you can't back away from the desk if you are on a roll; because tomorrow you might not remember the complete thought the same way it's in your head today.

The advantage of having the leg press built in the chair is not having to get up and go to another room and take your attention away from your work. Get your quadriceps built up while you are talking on the phone or doing data entry or any desk type employment.

HAVE WE BEEN VISITED BY ALIENS?

Let's get logical for just a moment. If we wanted to travel to another part of space, we would need to go fast... faster than a speeding bullet. Outer space is not completely empty. Our solar system is very old. It is clean compared to younger systems. Eventually, debris falls victim to gravity to become one with a larger body. Our planet is the result of debris collected together. This debris is all over space. When you look at a star on a clear night, it might appear to twinkle. This is debris passing between you and the star, somewhere in the multi trillion mile trip the light made to get to you.

In order to travel fast enough to get anywhere, we would have to move a vessel more than 1000 times faster than a speeding bullet. Light is about 750,000 times faster than a speeding bullet and it takes years to reach from one solar system to the next. Bullets go through things at gun speeds. Imagine moving the ship toward that bullet much, much faster. It would go right through the entire vessel. You could build a ram in the front of the ship 50 feet thick made of Kevlar, titanium, lead, and anything else you can come up with to stop a bullet and it would still go through it at the speeds needed to travel to the next closest solar system. The pieces of debris would corrode the ram away before your first year of travel. Even considering using lasers to break up debris before it contacted the ship... need I go on?

Let's come to terms with warp speed and folding space. Warp speed falls victim to the problem described above and nobody is folding space.

OK... OK. I'll humor the idea for a moment that some smarter life forms have figured out a method of travel. Now they've arrived. What are their intentions? Do they steal our precious resources and leave us? Do they wipe us out and claim the planet for their own? Do they hide on the dark side of the moon and watch us? Are they laughing? All realistic possibilities would include them making us aware of their presence. If they had good intentions and were smart enough to make the trip, they would be smart enough to avoid us.

We are barbaric, confrontational, and full of evil characteristics. We can't even get along with each other.

No. We have not been visited by aliens.

As for fast travel through space outside of our solar system, this will require a method of plowing the road, some sort of slipstream.

On earth, our atmosphere protects us from being hit by space debris. We have 11 miles of vapor to absorb the impact. Maybe we can plow the road by pushing a cloud of dense vapor, liquid, or both to absorb the impact of debris. We would have to work on a method of containment for either. I can imagine how thick this impact barrier would need to be in order to be effective.

TRY THIS

Go to sleep one night with your head lying on the inside of your girl's knee while she is lying on her back or slightly on one side, the side your head is on. You lay on your side facing her. Have her other leg over your waist. If one of you can't get comfortable then try a pillow between your head and the inside of her knee. To know you are in the right position, your belly should be touching her ham and buttocks. After you get very comfortable, try to fall asleep.

I INVENT MOST OF MY INVENTIONS DURING THE 30 TO 45 MINUTES AFTER GREAT SEX.

I think that this is because I am feeling satisfied and trouble free during this time frame. This allows me to think very clearly. My imagination is at its peek during this time. My heart has been pumping at a fast rate for several hours steady, sometimes all night long. This means that my brain has been fed lots of oxygen and nutrients from the extra blood flow during this sex time.

I had heard that Viagra use has a side affect. I heard it causes the brain to grow new brain cells and repair damaged areas. Now I realize that it's not the Viagra directly that does this. It's the longer sex sessions that Viagra allows for. Both partners have steady extra blood flow to the brain. The extra blood flow brings in the needed supplies for the brain to operate better. So, indirectly, Viagra and other medications like it can cause a person to get smarter.

We have seen this sex-genius pattern in several of our historical geniuses. Benjamin Franklin was a great inventor and genius and we know he was great at sex. He had an uncountable number of illegitimate children and the buzz in his neighborhood was that the women adored him. At the time it was believed that it was his self grooming and clean nature that attracted the girls but later evidence points to his ability to hand out sexual pleasure. I wouldn't doubt that he invented the majority of his inventions shortly after sex. Just the same, having an inventive nature allows a person to explore sex in new and more exciting ways. I can imagine that Franklin's charm was due to his inventive nature as is mine.

If you want to induce this affect on your self, your objective would be to feed yourself a few healthy meals in a row then raise your heart rate about 10 to 15% for several hours without depleting your blood stream of nutrients, oxygen, and chemical energy. This means you would have to use only a few of your muscles fairly steady at a light pace. A few small starving muscles will tell your brain that they need more oxygen. This will speed up your heart and breathing enough to flood your brain with extra resources. If you decide to use oral medication to raise your heart rate, you will need to be careful not to over do it. Cocaine and methamphetamines will raise your heart rate too much. Your heart can handle a faster rate if it is steady better then it can handle slow to fast and back again. You might be better off using a weight loss pill. Consult a nutritionist or a medical doctor to find out what will work best for you. Start out with a small dose on your first attempt and build your way up to the amount you can be comfortable with. This can become a method for students to cram for a test or a contestant to prepare for a game show. Your brain will simply work more efficiently.

THE COUNTERTOP SILVERWARE WASHER

The part about doing the dishes I dislike the most is doing silverware. It's tedious. When others in my household do the dishes, the silverware is not always clean. I'm over it! This unit will be a cross between a coffee maker and a jewelry cleaner with a few extra features. Imagine for a moment. You pour water in the top like a coffee maker and add a little soap. Hang the discharge hose over the edge of your sink and start the cycle. Come back in a half hour to clean and dry silverware. Yep, I plan to dry them too. The cleaning will be done by ultrasonic vibrations and water jets. It should be fairly quiet.

The real question is, do people want an appliance like this? It would be about the cost of a jewelry cleaner and a coffee maker combined. I came up with the idea a while back. The need for this appliance was greater then. I suppose it could still be desired by people who own dishwashers. This unit would be quieter and much more efficient then running a full size dishwasher. With increased use of disposable plates and bowls, the used silverware would build up without the rest of the dishwasher filling up. Also, you would have the increased convenience of where to put the silverware once it becomes time to wash it. Who wants to open the dishwasher door and pull out the big tray to drop in one spoon? It would be much more convenient to drop it in a counter top unit. I have friends who own a dishwasher and choose not to use it. To those who wash dishes by hand; don't you find the silverware to be tedious?

SLEEP VS. REST VS. DEATH

Step one: recognize that sleep and rest are different from each other and need to be addressed separately. After a few sets of tennis, do you fall asleep on the bench? Have you ever laid in bed rolling back and forth trying to fall asleep, but can't? You try to get up but you don't have the energy? These are situations where you need rest and you don't need sleep. I'm known to do this... Have you ever sat down to untie your shoes and found yourself waking up 15 minutes later? I've also been known to start a sentence and fall asleep before it's done. I don't have the sleep disorder, narcolepsy. I simply need sleep. These situations are rare because a person needs more rest than sleep. If we were to look at these two one at a time, we can see a difference.

Rest is like a cars engine. If you run it hard, you can overheat it causing it to shut down until cooled some. If you pace it, you can run for a while until you need more gas.

Sleep is more like your home computer. You can load in application after application until you notice the performance falling off a little. It just needs a reboot. Just the same, if you have a bunch of applications still running, it may take a while to shut down.

If I've worked hard all day at a job that takes little brain power, the drive home is my rest. Once home, I can start another project and stay at it for a while. If I've worked hard and used my brain hard, the drive home is my rest but I still need sleep. This is when I'll fall asleep while untying my shoes. Usually I will wake up about 15 to 20 minutes later feeling like I've got my second wind.

You are doing well when you figure out how to balance rest and sleep. For many people, rest and sleep feel like the same thing because the brain has the power to convince the body that it needs rest. This could be the cause of an overweight condition. If your brain is taxed enough that you need to sleep too long, you could try a 15 minute nap half way through your day and shave an hour off of your nights sleep. This would require some time adjustments until you fall asleep quickly and wake up refreshed. Here is another good tip. If you find yourself tossing and turning, get up. Do something. You're obviously not ready to be sleeping. If you have something on your mind and it's keeping you awake when you know you need the sleep, then start a movie that you've seen before. The sounds will guide your brain away from the thoughts that are keeping you awake.

What about vs. death? Why did I put that in the heading? I've noticed that the animals that sleep a lot have short lives and the ones that sleep the least tend to live longer. Also, another thing I've noticed is that cancer among other problems develops in the low blood flow areas. When we sleep, our blood is flowing slowest and sometimes we will lay on our arm, ear, or other part that will cause very low blood flow to the point where we will wake up to the feeling of pins and needles. That can't be healthy. How many hours have I been cutting off the blood flow? What about sleep apnea? A little less oxygen never hurt any one, right?

YOU CAN'T COMPREHEND IT UNLESS YOU DO IT YOURSELF

When I first wrote this article my intentions were to show how our communication methods lack the ability of comprehension when the feeling is something the audience has never experienced before. After reading the article back to myself I realized that I got caught up in the story and never did get to the point. I almost cut this article from the book but then I decided to leave it in because I think it's entertaining.

Several years back, I was just hanging out at my friends place when another friend arrived. He was riding a motorcycle that he had just built. The new arrival gave the keys to my friend and told him to take it for a ride. He did. When that was done, they both tried to convince me to take the bike for a ride. My friend was excited from his test ride. He said things like "You gotta try it!" "You have to ride this bike!" I refused on the grounds that I don't like to be responsible for other people's toys.

I owned several motorcycles at the time and I didn't like other people to ride them. My motorcycles were not to be laughed at. I could be doing 200MPH within 1 mile on my best bike. How is someone else's bike going to impress me? I was an extremist at the time. I took risks for the adrenaline rush. I had raced on the high banks of Daytona and I had been known to ride wheelies at speeds more than 140MPH and hold them for several miles at a time.

After about 10 minutes of badgering with things like "You'll regret it if you pass up this ride" and "I can't explain the feeling. You have to feel it for yourself!" I finally broke down. The curiosity got the best of me. I agreed to try it out.

I pulled away fairly slow to get a feel for it. The road was straight for a quarter mile then curved left gently then back to straight for half mile or more. About half way to the curve, I pulled a little throttle. My body was fully tucked down with my arms bent and hanging on tight. I felt the passenger seat hit my butt and the bike pulled very hard. I backed off for the curve and took it slowly. When I straightened out, I shifted to third and pulled the throttle all the way to the stop.

This bike was very long and low. It was built in a way that it was good for drag racing and not easy to wheelie.

The bike pulled harder then before. If the passenger seat wasn't there, the bike would have pulled itself right out from underneath of me.

It was all that I could do to hold on but I was still in control. I wasn't impressed yet. I had my thumb floating over the boost button but I hadn't pressed it yet. I figured I was there to feel what they were talking about so I had better use the button. All right, I'm just going to touch it for a second.

BAM! The power came on like a Mack truck had hit me from behind. The entire bike came way up on one wheel with a surge of power that obviously doesn't belong to a motorcycle. I had already let go of the button but the bike didn't care, we were moving forward intensely! I'm just glad the bike didn't flip over. I know that I only touched the button for a half second at best. Now, I'm impressed! I think I started yelling something to myself in the helmet. I know I was smiling from ear to ear. It was one of those situations where you can't stop yourself from smiling for the next thirty minutes. Remembering back to that moment still makes me smile. Later I found out that the 400LB bike had more horsepower then most 3000LB sports cars have today.

I know that I can't possibly explain the feeling to you. I can use all the words in the dictionary and you won't comprehend what I felt. You can ride the amusement park thrill rides and you won't get 10% of it. You will have to do it for yourself to comprehend. I've heard it said that life is just a series of moments. I think that this is what they were referring to.

No Guess Microwave Oven

Microwave oven wattage varies from 500 to 1800 for standard home models. It is because of this that food cooking instructions are not precise. Every time you place something in to your microwave you have to guess the proper amount of time and power level. Guessing leads to burning the roof of your mouth or having to put the food back in the microwave to warm it more.

The technology of the basic microwave isn't my invention. My invention is the sensors used to tell the microwave how much heat to apply to the food to reach the desired temperature.

The hot foods we eat are usually around 115 degrees and the hot drinks we drink are usually around 150 degrees. This will become common knowledge as people fine tune the temperatures of their favorite foods and drinks. Essentially, a microwave oven could have just a few buttons; thaw, warm, hot, and boiling. The microwave will sound its finished beep when the sensors determine that the contents have reached the proper temperature. Also, in the same fashion that a computer will show you estimated time remaining, the microwave can do the same. It can do a count down to zero.

A person would put in a sandwich and press warm or a cup of coffee and press hot. Other models could still have the same keypad as they do today. A person would place a slice of pizza inside and key in 120 degrees. When you hear the beep you know that you can jam that slice of pizza deep into your mouth without burning yourself.

The whole point of the microwave is to be a time saver. Why wait for it to overheat your food and then have to wait for the food to cool before you can eat it? Also, why waste the extra energy?

Have sex in broad daylight

This falls under the category of being rich. Many people don't know that it doesn't cost much to live rich. Most believe that money makes a person rich. I know better. A person lives rich when they reach a particular level of comfort. Many comforts are cheep or even free. Yes, some comforts are expensive and out of my price range. I will strive to reach the comforts I can afford. We can't afford a private beach but we can afford to make our yard so private that we can have sex outside in broad daylight if we choose.

To be rich you need to eliminate stress and achieve a level of comfort where you feel like you can do anything you please. This means you have to have the proper environment and all of the little things that please all of your needs. You will need tools, shelter, toys, foods, drinks, and company. The difference between living rich and living poor is having everything you need and not necessarily everything you want.

You can design a comfort zone. Acquire a spacious area where you will be able to do with as you please. Your yard around your house will do just fine if you live in warmer climate. If that's not possible, then rent or lease a warehouse or team up with a close friend who has a yard. Just do what you have to in order to set the stage. If you want to use your yard, you will need to make it very difficult to see inside. Get your yard ready by bringing in dirt to make a 2 to 3 foot tall mound around the perimeter of your property. Shape the hill to be gently tapered on the inside and have a sharp drop on the outside. Then install privacy fence on top of this mound near the sharp drop. I'm not talking about the kind of fence you can walk up to a crack to see through. I'm talking about the kind where each piece of wood overlaps and you can't see through no matter what. You may have to putty in a few knot holes.

If you already have privacy fence, pile a 2 to 3 foot high mound of dirt against the inside all the way around and plant grass seed in it. Landscape it so it is like a rolling hill and not a sudden steep incline. Yes, that will take a lot of dirt. If you don't have time to wait for grass to grow, bring in sod. After a few weeks, allowing the dirt to settle in and harden, remove the fence panels and raise them up by a foot or two and re-secure them. You may need to scab an extension to the top of your fence posts, but I'm sure you will figure out how to do this. Make your gate remote controlled. My gate is on wheels on tracks and opens sideways. It is simply opened and closed by a garage door opener. I made it line up with the rest of my fence when closed so that a person can't tell where the fence ends and the gate begins. Upon opening, the gate pulls out and sideways. I just angled the rails.

If your neighbors have a tall house with windows overlooking your new tall fence, you will have to relocate or bring in shrubs or trees in order to block the view from the windows. I hung a long container full of potting soil on the inside top edge of my fence and planted some self maintaining plants.

Now you can walk around your yard naked and not worry about being seen. You can be comfortable. If you are still uncomfortable, you can mount motion detectors, microwave sensors, or shock sensors around the perimeter to warn you when somebody is invading your privacy. They would be breaking the law by looking over your fence.

In order to make your yard inviting to friends and enjoyable to all, you will have to buy some toys. Hot tub, trampoline, a large seesaw, a large high quality swing set, heated pool, screen house with a couch, rocker benches, nice lounge chairs, picnic table, gazebo, water slide, several hammocks, lots of patio furniture, rope tire swing, monkey bars, tree house/clubhouse, jogging or walking path, Chin-up bar, inclined sit-up bench, dumbbells and work out benches, an extra large decorative yard fountain and waterfall… You get the point! Just buy the items you can afford. If you are low on cash, you can make several of these things yourself. You can make a seesaw, swing set, chin up bar, etc. Just make them large and use quality materials. You can find some of these toys used. I bought my hot tub on the internet for $450. It's big enough to seat 6 and works great. I had a hard time picking it up but it was worth the trouble. I connected bungee chords to pulleys at the tops of two of my palm trees and placed my trampoline between them. With the harness on, a person can jump almost 30 feet high. It's a whole lot of fun for the money.

Between buying the fence and a bunch of toys, you may spend a sizable chunk of money but you don't need to be wealthy. The average home owner can do this on one credit card.

You should alleviate the stressful parts of daily life. Buy a bunch of fingernail clippers, scissors, pens, pencils, paper, quality flashlights, calculators, small radios, candles, lighters, new socks and underwear, hair combs and brushes, extension cords, telephone extensions, notepads, copies to your keys, extra remotes for your garage and car, outlet splitters or power strips, blank digital media, a high quality digital camera and the largest memory it will take, desk organizers, candy bowls, universal TV remotes, ant & roach sprays, weed kill, hand tools, batteries of all sizes, personal lube, condoms, towels, sunglasses, and anything else you would normally stress about when you can't find it. Develop a system where you will write down any of these things on a shopping list every time you run low or simply can't find the item in the first place you look. Buying a bunch of these little things isn't as expensive as people think. You can get many of these things at a dollar store. The stress of not having something when you need it is worth a whole bag of trinkets. Think for a moment how expensive a special trip to the store costs with gas prices today and what about the wasted hour needed to go get another battery or can of bug spray because you need it now.

Work with the other members of your household. I kept buying tape measures every time I couldn't find the last one, only to find out that my girl was putting them somewhere that I wouldn't check when I needed one. Since I buy 3 or 4 at a time from the dollar store, I wouldn't complain but once in a while. We must have had 15 of them before I complained to her and she showed me where she was hiding them.

Now you are living like you are rich. You can hold private parties for your closest friends. Remember that company and friendship is an important part of being rich.

You can also modify your attitude. You're rich now. Don't let anything less than $10 cause an argument. Would you pay $10 to make your life less stressful? So what if you were charged double for your case of soda. Is it worth waiting for the manager and dealing with a possible argument? Just eat it! You don't want to be called cheap behind your back. $10 is not that much money. Move on with your life. You have an awesome lifestyle you need to get to.

A few days ago and on a regular basis, my father tells me a story how my step sister used his flashlight or something else that he needed, and didn't return it. He gets so mad and stressed out over something that costs a buck! The health problems related to stress will cost more than a buck. This is my trick; I buy the quality flashlight and a 3 pack of super cheap flashlights at the same time. I put the cheap flashlights in front of the quality flashlight on the shelf where we expect to find them. Others in my household will grab the most convenient flashlight in the front. When I find it in the bottom of the pool, I don't get mad because I know it was only a buck.

I DON'T THINK TOILET PAPER IS GOOD ENOUGH

Sometimes I wipe and come back clean. Other times I wipe 5 times and still feel like I'm not clean yet. These are the times I think I could use a something else to do a better job. I'm thinking of a finger wrap that you can drop in the toilet.

It has been suggested to me to use baby wipes. This is when I realized that I was being misunderstood. My intentions are to make something that will clean up inside a little. If I've wiped 5 times and still keep coming back with some on the wipe, then I must have some stuck in limbo, close to the exit without coming out. This is what I want to be able to clean out.

I have been told that it is customary to use a finger through the toilet paper in another country. I figured the guy who told me was pulling my leg. If what he says is the truth, then that country should already have something like I am describing.

We have some materials that melt in our stomachs like medicine capsules. I believe this material should melt in our sewer systems and be safe to flush as well as be safe to introduce into our colon as a type of internal wipe. My cat got a hold of a capsule and chewed it for at least 15 seconds before I got it out of her mouth. The capsule material used is strong! It was mangled but not broken open. It has to be moderately flexible too.

I think it would catch on if some studies were done or we got some doctor recommendations stating that it would reduce the chance of colon cancer. (It might reduce the smell from passing gas???)

GIRLS… DON'T MAKE THIS MISTAKE

My ex-girlfriend is a very sexy girl. At one point, she packed on a few pounds. I didn't mind the extra weight. The readout on the scale troubled her so much that she began to wear clothes to hide her body. I didn't like the new clothes. I wanted to see her hour glass figure. She misunderstood my reaction to the clothes and figured I didn't like the extra weight.

My current girl is very sexy. I told her I wanted her to put on a few pounds because it makes her more voluptuous. She agreed. She put on ten pounds then cut me off. A week or two later she's back to her normal size. I think she would also be ashamed of herself if she couldn't lose the weight.

My ex-girl never once stopped being sexy. Her personal shame caused us lots of grief. If my present girl stopped showing me her sexy figure, I would probably lose some of my sex drive. She might mistake that for a feeling of disgust.

PUT YOURSELF IN THIS POSITION

At a concert I watched a guy standing in front of me slowly fade away, getting higher and higher until he could no longer stand at the railing. After an hour, he couldn't support his own weight using his legs and arms together. He was now sitting in a puddle, leaning against the lower half of the rail which was made of cement. A few people approached him and made suggestions for him to lie on the lawn behind the seats in the back. Others tried to help him up. He was too limber to be lifted by one drunk at a time. Most people wanted to help him but didn't want to get soaking wet now that he was lying in the large puddle. A friend of his tried to pick him up for real. This guy sacrificed his nice clothes and really struggled to get his friend moved.

Now the guy is so faded away that he jumps suddenly and wiggles his way out of his friends arms back to the ground. The way the friend was carrying him must have been hurting him and it took him a few seconds to become aware of it. His only thought was to stop the pain by getting out of the grip on him. The ground was much more comfortable, even in the puddle.

What was he thinking...? Wait, I can't get to the lawn myself, I need your help to get there. Hey, don't leave me here in the middle of the walkway. Why can't I speak? Why can't I get up off the cement walkway? Why am I all wet? Oh, who cares? I feel really good right now. I'll enjoy the feeling for just a minute and I'll close my eyes just for a second. Whoa, what the hell? Why does my arm hurt so bad? DAMM, that really hurts. What the hell has a hold on my arm? Awe, DAMM! It's a cop! Hey guys! I'm OK. No really I'm all right. Why can't I speak? Awe man! I'm too high to speak. This really sucks! What can I do? What can I do? Nothing! Ouch that handcuff is pinching me and I can't get my weight off of it. How the hell did I end up like this? I'm too stoned to remember. Now I have to wait until this horrible high wears off. I'm just screwed!

Why didn't I help him up to the grass? In the back of my mind, I kept thinking to myself "What if he is poisoned?" If I helped him up to the lawn, he would get no attention. He could convulse and die without anyone noticing. If I leave him up to the police, they will watch him enough that if that happened, he would get medical attention in time enough to save his life. I feel bad for allowing him to get arrested. I could have helped him to the grass and he probably would have been fine. No trouble would have come to me except for some wet clothes and a little effort to carry him there. I would have wanted some help if I were in his shoes. The help I would have wanted would be for someone to help me to the grass and watch over me incase I did start to convulse or choke on my own vomit. That is a lot to ask of someone. That is what separates a friend from an acquaintance.

I LOST 50 LBS IN 50 DAYS

Several years ago, I looked in a mirror and felt it was time to loose some weight. I was almost 230LBS and I felt I should be about 180LBS. My extra weight was evenly distributed and not noticeable in any one area. I was putting on about 10LBS a year for the last 5 years. At 220LBS the year before, I still didn't feel fat and I was still in decent shape. I was just thick.

Over the next 50 days, I lost a pound a day. Before people asked me how, I really didn't think about it. I just knew that I wanted to be healthier. Each time I ate, I would remind myself that I wanted to be healthier. It became second nature within one week.

I did add some more activity to my day. I ran an automotive business and therefore didn't have a lot of spare time. To add more exercise, I began moving quicker. I would walk fast from place to place or even run out to the parking lot when I needed to get a VIN number or pull a car in to a bay. I would choose to get up and do something myself even if I had an employee available to do it for me.

Once I noticed some serious weight loss about 2 weeks in to it, I became enthusiastic about becoming healthier. I began lifting weights 3 or 4 times a week. At this point I was feeling really good about myself and that gave me the energy to keep at it.

People started noticing me getting thinner. One friend of mine, Tiny, was big fellow. He was always big as long as any of us had known him. I didn't ask him how much he weighed but I'm guessing it was 400LBS or more. Tiny noticed that I lost weight fast and asked me how I did it. I told him how and then he proceeded to lose 150LBS in 150 days, roughly.

My system has worked well now for several more people. It's simple to do. You might want to give it a try. If you eat at night, rule 1 can be tough for about a week. The other two rules are easy to do.

Rule 1: go to bed hungry.

Dinner at 6:30 was the last time I would eat.

Rule 2: Keep the stomach size small

Eat smaller meals. Eat more meals if you have to. I didn't change the types of foods I ate. I kept eating the pizzas, subs, Mac & cheese, burgers, etc. I would eat smaller amounts.

Rule 3: Flavor control

When eating foods, drink water. Only drink flavored drinks by themselves. This is because you may begin to eat a little healthier and these foods have less taste. You can be more satisfied with less taste if you compare the food taste to water. The food flavor would be drowned by chocolate milk or soda. You can never convince yourself diet soda or other diet drinks with flavor are OK because they have no calories. It has flavor and that will raise your flavor expectations on other foods. As a rule of thumb, more flavor means more weight. If you do drink weight loss shakes, drink them by themselves when you know you won't be eating any time soon. These shakes can backfire if you eat foods with them.

If a person is serious about loosing weight, it is very easy to keep these three rules on their mind. The following are a few tips to assist through the hard parts.

If starvation is too much to bear, try drinking a lot of water before bed so that you have something in your stomach. Try to stay busy. A person thinks about food more when they have spare time. The first week will be the toughest to deal with the hunger and general changes to your habits. Just know that the third day is the hardest. It becomes easier from that point on so hang in there! Don't give in on your 4th or 5th day because it becomes easy after the first week. Once people start telling you how good you look, a new kind of proud self image becomes your motivation to keep going and it becomes very easy to stay strict to the rules.

GIRLS... DO YOU WANT TO KEEP YOUR MAN?

I'll give you an idea that will help. You will need to come to an understanding about the male brain. It is nearly impossible for a man to completely ignore other girls. I'm not saying we are doing anything wrong. I'm saying we are noticing others. You could be the hottest girl on the planet and your man will still look at other girls. Variety is the spice of life. Will your man be with other girls? I can't say for sure. I can say he will consider it.

With that in mind, ask yourself, what can I do to reduce the chance of him acting on it? You can offer him variety in yourself. If you provide some variety at home, he won't look out the window as much. You could try several hair styles and colors. You can try on a new attitude for a day. Get into some art for a minute. Show interest in his work or hobby. Smile. Glow. It's the most attractive feature you have. Dress up just for him. If you only dress up when you are going out, it will seem as if you are shopping for another man to him.

I don't know if girls consider being with other guys as much as guys think about other girls. Logic says yes but we all know that logic doesn't fit very well in love and relationships. If you notice your partner looking at other people, maybe you should consider doing some of the things I described above. We should offer variety in ourselves to each other. It will make all of us happier.

WATER WALKER

Have you ever wanted to take a walk across a lake? Look at the benefits, you can stay dry and you don't have to use your back muscles to row. You face the direction you are traveling. I'm not knocking rowing but sometimes I want to stand up in the boat or use my legs to row with and I would prefer to travel forward. It's just easier to see where I'm going. The walker is a real simple machine. It is a square basin with Styrofoam edges and two handles in the front that lead down to two additional floats that are used for stability and navigation. The inside floor is tilted up in the front with a conveyer belt that you walk on. The conveyer belt is on rollers so it will move easily and it is attached to an output shaft connected to an impeller underneath. Your walking provides the energy to turn the impeller.

This will give a new definition to "sea legs". I see no reason why you can't walk across a lake and stay dry. I would make it out of plastic, Styrofoam, and aluminum for the handles and rollers. I would build it to be so light that one person could easily carry it from the trunk of the car to the water and back again. There's no need to make it big enough for two people, each person would use there own. You could walk side by side on the water by latching two of them together. It would be much more stable like that. A new comer would have a better chance of developing a sense of balance on the water while attached to another person. I think I could make it very cost effective if I made a bunch of them, $50 give or take a few bucks.

GIRLS HAVE SEVERAL MAGIC POWERS

The girls with the most magic don't even know it. The girls that use it the most don't use it the right way. The girls with the least magic will learn the most from this article. You don't believe me? We'll run some tests. We'll need a girl. If you are a girl then try the following without letting anybody know what you are up to. I don't want tampered results. If you are a guy, get a girl to help you run these tests. Same goes, only you and the test subject can know what is going on. All of a girl's power comes from drawing energy from other people around her. If she were to draw too much power too fast, some of the people will feel this and get angry. Powers should be used in moderation.

First Power; Telekinesis or Telenergy: the ability to move objects by means of thought alone. This power comes with a side affect of momentary bewilderment of those who watch. This power is used in very short bursts and can't be used to lift anything more than 10 pounds. The probability for success is better than 50% for the average girl. This power will be stronger the second and third times it is used in succession, but its strength will taper off from there. The best working distance is from 2 to 5 feet. A good test environment would be at a wedding reception or a job Christmas party, but not a family gathering as much. Be sitting with a three or more person group of at least 50% men.

Test procedure: first you must choose the object you wish to lift. It should be small enough to be picked up with one hand and it should be out in the open. You don't want to pick an empty glass in a crowd of other glasses. The object should be obvious to others who glance at it quickly. A good object to move would be your own drink. You can set this up as you walk over and sit down. Deliberately place your glass out of your reach.

There are two ways to invoke this power. In either case, once you start the process, you may not speak of the object at all. One word about the object or somebody else touching it or placing another object in your way or any word at all would dissipate all powers for about 15 minutes. If someone moves it or places their glass next to it, just consider it more of a challenge.

First way; while speaking to someone about whatever is the going subject; reach your hand out toward the drink as if to shake its hand. Have your fingers spread apart. Your hand should look like you are holding an invisible football with the balls pointed tips facing up and down. Do this directly toward the object you want to move. Remember that you may not speak of the object at all. If somebody else says anything at all about your arm or the object, you may act like you are bouncing the invisible football against an invisible wall 1 inch in front of it. That should be enough to conjure the nearby energy to move the object toward your hand. Pay no attention to the object with your eyes. A quick glance should be enough for you to line up your arm with it. If you are in middle of a conversation with a nearby person, you will be able to easily follow all of the rules in order to see the magic. Remember not to speak of it. If you must say something, you may say "Thank you."

Second way; when you are not speaking, you can stare at your object. Lean your head a little forward and concentrate on moving that object. Again, the "no talk about it" rule applies. If someone else barges in and breaks your concentration, you can do one of two things. You can fall back on the first way and just lift your arm as described above or you may say "Please." Remember not to speak of it. If you must say something, you may say "Thank you."

I know this sounds like a simple case of using other people. It's more than that. I sat next to a girl who did the first way right in front of me. I knew I felt something unusual as I watched. A few minutes later she leaned over to me and said "Did you see that?" We talked about it for a moment. It turns out we both felt an energy from this that neither of us had felt before. Maybe you will understand that the power isn't the moving of the drink, it's the way you affect the feelings of those around you. It's as if you are massaging the feelings of those around you. You give the feeling of acceptable confusion.

Second power; Spell Casting: the ability to cause another to act in accordance to reach your goal. This is a very strong power and can have devastating affects if used excessively. This power has been known to kill if the goal is out of reach. Please use it wisely and fairly. The probability for success for the average girl is about 50% with male victims and 20% with female victims. The power is strongest when the victim doesn't know you and it is weakest when they know you very well. The working distance is best with direct contact to victim. The chance of success drops dramatically with repeated use on one victim. Using this power in a group on several people at the same time will create chaos and disorder. Just call the police as soon as you notice a girl casting multiple spells. This spell strength can be adjusted with internal energy. The more energy released at the time of casting, the greater the affect will be. My suggestion is to start with very small jobs to give to your victims. You want to get a feel for how the spell works before you put someone on an unstoppable mission.

How to cast this spell; politely ask the victim to do the job you wish to have done. Wait one moment. If the victim agrees then you did not need to cast a spell at all. If the victim hesitates or starts making excuses then a spell will be required. If the victim is putting up a fuss, you will have to put a lot of your own energy into the casting of the spell. If not, a basic kiss on the cheek will cast the spell just fine. If more energy is needed, you can hold one hand and give a big smile after giving a kiss on the cheek. If it looks like your spell has failed, just walk away quickly. Remember, the average girl has only a 50% chance of success.

If you are a girl and you think this is hogwash, then go back to read the first sentence of this article. If the first sentence doesn't apply to you then maybe the third does. In either case, you have to try to see proof.

I am going to stop here for now. I hope you can see the energy that I am describing above. The average person has a soft spot for cute girls. This soft spot includes a feeling that is a natural protective instinct. I would help a damsel in distress before I would help a grungy dude with the same problem. I feel like the guy should be more competent to help himself and I wouldn't expect the girl to be as much. I would help the girl first. This energy is in most people and it can be tapped in to with a little persuading. Have fun with it but don't abuse it.

ATM RECYCLER

I will design a machine that will look like an ATM machine and does the work of a recycler. You can drop in a bunch of used plastic bottles and get out a laundry basket or a desk organizer. Maybe you will drop in about 50 soda cans and get out an oven baking pan, some clothes hangers, or some shelf brackets. The machine will melt down the products you put in and have a menu of products that can be taken out of it. You don't have to take something right away. The machine will store your information and your deposits until you need something. All machines will be linked so that you may use your credit even when you are traveling. Imagine being able to use a laundry basket and clothes hangers while you are vacationing and just redeposit them in to the ATM when you are done to get your credits back. As this machine evolves, we will give it more and more products on its menu for you to choose from. I can imagine that eventually the machine will be able to accept all forms of trash. It will decide what is useable and what should go to the dump.

THE ONLY FOOD LEFT IS ROTTEN

I received a call today from a friend who lives about 4 hours drive away from me. She said they haven't had electricity for more than 3 days now and the local gas stations are out of fuel. It is completely dark at nights and the looters are smashing and taking everything of value. The grocery stores are wide open for the taking but the only food left is rotten.

Suddenly, my problems are unimportant. I wouldn't have known that it was that bad if she hadn't called me. It doesn't hit home the same from the news stations as it does when your friend calls you.

This proves that we are too dependant on power company electricity. If the food wasn't rotting, they wouldn't be in a rush to get out of there.

This is the result of a hurricane. The hurricane is the result of global warming. Global warming is the result of humans burning fossil fuels. How much longer will we give ourselves?

HAVE YOU EVER STOPPED TO THINK ABOUT WHAT HAPPENS IN A NUCLEAR EXPLOSION?

If I paint the picture for you, then you will see the importance of knowing. A nuclear explosion can occur several different ways. The one we hear about is the one generated by mixing plutonium and enriched uranium. This is the combination used in the majority of our nuclear devices. These 2 elements can break down one another. When the plutonium is exposed to the uranium, an exchange of electrons begins superheating the atoms. These super hot atoms begin to break down. A portion of the nucleus of each atom breaks off to form a new atom. Most of these newly formed atoms will be from the small end of the elemental chart. Lots of hydrogen, helium, Lithium, Beryllium, and occasionally some other elements will form.

As will happen many times instantly, a helium atom will break off of a uranium atom. The result would be two atoms where only one existed one billionth of a second ago. This keeps happening until the large nucleus breaks down to several smaller atoms. The resulting Thorium is still radioactive and superheated, therefore it will continue to break down to become Radium or Actinium and another atom will be formed. This will continue until the uranium becomes lead. Lead is 10 protons and 21 neutrons less than uranium. These 31 particles will form the smaller atoms mentioned above. Most of these atoms will be improperly balanced. They will have too many neutrons and too many electrons. They will not mix well with the atoms already existing here. They will beat up existing compounds, materials, and organisms for the next 30 years. This is the truly awful damage done by a nuclear weapon. We are **incapable** of cleaning up after it.

The problem with the 2 to 10 new atoms formed is that they are going to need space to exist. This breakdown or fission is happening to billions of atoms all in an instant. This means that in a billionth of a second, a couple hundred billion, no... a couple hundred trillion new atoms are going to demand space to exist. What I am describing

---

is an extremely high pressure space where the new atoms are pushing outward against all particles adjacent to them. This expansion happens instantly, crushing and heating the atoms nearby to super high temperatures. This is the smaller explosion that occurs first. Actually, this explosion is very small compared to the next explosion to follow. Even though this explosion is small, the immediate surroundings will disintegrate when hit with atoms moving at the speed of light.

A type of shock wave will be generated that will carry through all adjacent solid objects. The ground will carry this wave for a long distance. Imagine the ground looking like water in a still pond where a pebble just broke the surface. The wave will carry outward from the center with a large wave forming the largest circle and several smaller waves following closely behind. This will happen a lot faster than what you see on a pond. Figure the pond's wave is 1:100000 scale. Watch the water's wave spread and imagine the ground's wave doing the same. The wave will spread at about a mile per second. If you feel this wave you will have enough time to consider yourself dead because the next explosion will be much larger and more violent. The truth is that you won't feel this first wave. It will smear everything together like a huge steam roller. The first ¼ mile will become a murky paste of everything solid just powdered and mixed in to the liquids. The next few miles will appear to be 10,000 years older instantly. Buildings will crumble and anything moist will evaporate almost instantly. If you were standing about 5 miles out, the ground under you would snap upward and shatter both your legs probably up to your waist.

Above, I described the uranium becoming lead. This is a very temporary state for the matter. Some of the matter will turn to gold. You don't want any of this gold because it's not 24 karat. It might be 24K degrees. The extremely high temperatures will also cause further break down of the atoms. Any kind of atom can form at this point. All of these newly formed atoms will be hot and under high pressure. They will all require space to exist. They will push outward with the kind of pressure that will smear you down until you are only 1 inch thick. Many of these atoms are volatile when mixed. The high heat will start a sphere of fire around the original reaction point that will be pushed outward so fast that most of it will burn out like a candle being blown out. For a fraction of a second, the high pressure will drop most of the way to normal before the remaining atoms begin to react with one another to relight the fire sphere. This time the sphere will remain on fire, building pressure again and pushing outward from the center much slower than the first but still very fast compared to how fast you can run. The top and bottom of this sphere of fire will burn itself out quickly from the pressure changes but the sides will form a ring of fire that will move like a huge wave of water outward from the center and a little faster than water would flow.

If a spoonful of each of these elements was used, the destruction would span for several miles. If a cupful of each mixed, the destruction would span for 50 to 100 miles. I'm not sure but I think the average bomb has about 4 cupfuls of each in it. Don't even try to do the math. Just know that these elements can be used for much better purposes then to threaten other people.

Nuclear waste is an issue. We have power plants that use these elements to make electricity or to power submarines. What do they do with the waste? I believe it is possible to use these elements in such a fashion that they will leave no waste at all. Doing so will require the same kind of effort we put into our cars to make our exhaust cleaner.

It would make sense that breaking down elements to form new ones could be done in a fashion that the final elements don't need to be unbalanced or radioactive at all. Do the existing power plants do this? I don't know. If they don't, then they would be producing the elements on the chart just above lead. These elements are radioactive. The rest of the new elements would be from the small end of the elemental chart and will be unbalanced. We can't let these unbalanced atoms mingle with our stable chemicals. In my opinion, if we can't break down these elements clean, then we are not ready to use them to make energy.

ARE YOU PICKING YOUR NOSE AGAIN? MAYBE YOU SHOULD BE.

I'm not a doctor. I'm not even a nurse. This doesn't mean I can't point out patterns in my health. I admit it, I pick my nose. I've got a pretty good technique down. I use the little finger on my left hand. I'm very clean about it and I only do it in private. I wash my hands between 20 and 80 times a day. Some would say I'm borderline obsessive compulsive. I'm NOT! Thank you.

For my technique to work well, I need a little bit of the finger nail to be grown past the contact point with the skin. Some times my nail breaks off even with my skin or worse, below the contact point. It may take a week or two before I have enough nail grown back to get back on top of my game. During these times, I'm more likely to catch a cold.

I'm familiar with the rhinovirus. He looks like a miniature version of a rhinoceros. He clings to the inside of your nostril and causes you to create mucus. The mucus flows out your nose and down your throat causing you to sneeze and cough until you have a sore throat and liquid in your lungs. This leads to a lung, sinus or throat infection and you're sick! I'm good at scraping these guys off of the inside of my nostrils. I'm very healthy. I rarely get sick. When I do get sick, it's usually when my fingernail is broken off! If you don't want to pick your nose, you can use a spray that does the same effect.

THE WEATHER AFFECTS MY SEX DRIVE

I've noticed my sex drive falling off lately. I feel wore out after working a long sweaty day in the heat. I can tell I don't have what it takes to offer my A game or even my B or C game. It hits me the hardest at the end of a 6 day work week. If I'm worn out from work, I don't want to go back to work in bed. I know the weather is a large part of my lost sex drive but today I discovered another part of my lost sex drive. I lose it when I'm not satisfied in other key areas. I need to feel accomplishment. I feel energized when I see something I've worked at come together. When nothing works out for me, I tend to feel a little depressed.

I need to be able to masturbate regularly. I've been telling my girl since I met her that sex and masturbation are 2 completely different things and they must be dealt with separately. One does not replace the other even if done excessively. If you are a man then you know what I'm talking about. When you didn't have a girlfriend, it wouldn't matter how often you masturbated, you still needed a girl. When your girl is new, you are too busy to miss it. After a while, the need comes back. This comes from a man who had over 800 hours of sex with his girlfriend last year.

My girl was reading from one of her girly magazines last week when she approached me with an article that said the average man has 16 hours of sex each year. She chuckled then said "We have more sex then that on a single weekend!" I thought about it for a moment. If a guy lasts 15 minutes at a time and does this about 5 times a month, he would have about 16 hours of sex each year. Later, I sat down with a calculator to figure our sexual time together last year. We averaged about 17 hours of sex each week with our longest stretch being a little more than 2 days straight. This is a good day and age now that erection enhancement supplements have been made available.

I like sex the most when it starts up naturally. I don't like being told "I'm horny… so let's go play." That is a turn off and it puts a little pressure on me to perform at the drop of a hat. It's obvious that her sex drive is greater than mine. I'm not a school boy anymore.

Another thing I've been telling her since we met is that my sex drive is like a huge pendulum of a grandfather clock. It is very slow to swing from one side to another. I get into sex for a while then I've had enough for a while. The time it takes to swing from max on to max off is about 2 weeks. During the time I'm not into it, I feel burdened by her need for it. During the time that I am into it, she gets all she can and still can't get enough.

My friend once told me that behind every great man is a great woman. A strong sex drive is a key element to a great woman. Lately I've noticed her patience wearing thin on the weeks that my sex drive is low. I've also been telling her that girls who masturbate regularly can achieve orgasms faster and more often then girls who don't. I would prefer for her to masturbate during the times my sex drive is at its lowest. Actually, she might. I think she is too embarrassed to tell me things like that. She is sensitive and shy, not like me at all. Last night, she asked me if there was anything in this book that she might not like. I know she isn't going to like all of these things I tell about her, so please don't throw any of it in her face.

WHY HAVEN'T WE REDESIGNED OUR BEDS?

The bed hasn't changed in centuries. I'm not talking about replacing springs with foam. I'm talking about the shape of the sleep surface. We should shape the foam to conform to our bodies. I want to lay face down with my arms out to the sides with my hands near the top of my head. My elbows would be bent about 90 degrees and my arms would be below my body. I would look like I were half way of a pushup. I would need an opening for my face and another for my family jewels. I want my legs to be bent so that they are lower than my back. I would look like I was in a stretched out crawl.

I also want a side position. Many people sleep on their side while gripping a large pillow. We should contour the mattress to better fit our lower shoulder and make the pillow have groves for the upper leg and arm and be designed so that it wouldn't be pressing on the lower leg and arm. I like to sleep with my chest down, one knee out to the side, one arm out to the side, and my other arm under my pillow. I would like the mattress to have an opening for my lower arm to go under my pillow comfortably and a grove for the hip and quad muscle to dip into.

WE WERE GIVEN THE TOOLS

I seem to be from the last generation of men to be brought up tough. I had strict rules and I received the needed spankings to teach me how to be strong, aggressive, and hard-hitting. I was pushed hard until I learned tolerance, dedication, perseverance, determination, and patience. I was required to handle and tolerate pain. I can handle a wide range of emotions so well; I can trick you into believing I am autistic. Yea, I can hide all my emotions under any circumstance. No problem. These are the tools needed to be tough.

My girlfriend's son has 2 pencil tips stuck under his skin. Yesterday he got a splinter under his fingernail. He won't let us get any of them out. He cries like a baby if we mention knife or needle. I suggested to my girlfriend that we hold him still while we remove the items. The problem here is that she doesn't have the ability to listen to him cry. Her weakness is being transferred to her son by example. In my opinion, he is growing up to be a baby girl.

As time passes, I realize that a new kind of evolution is taking place right now. When genetic mutations occur, the stronger strain will live on after the weaker die off. This is the old kind of evolution. The new kind is more based on mental competence. The stronger we are mentally, the more likely we are to die off. No, that's not a misprint. The educated are more likely to die off sooner than the less educated. The more knowledgeable individuals are tied up with responsibilities that make it difficult to fit in time for family. I am in my mid 30's and I'm beginning to think that I don't need to have children. My friends who are business professionals tell me that they don't want to have children. Am I the only one who sees this as being a problem?

I'm off the subject I started with. A man born of my generation has a hardness to him. He is stubborn in his ways because he feels like the trouble he went through as a child was for something more then what he added up to be. He will not be accepting to new concepts. He will put up a fight to no end to uphold his traditional concepts and beliefs. This man is of my own kind and he will be my worst nightmare when it comes to my introduction of new methods for better living. The men who grow up to be baby girls will accept my concepts without a fuss.

Most of the men who were pushed to be tough are from before my time. Many of them feel like they were left behind or didn't get a fair shake. Many of them hold a grudge against all who are younger. They know somewhere in the back of their minds that life could have and should have been better for them. They feel like now it is too late to make the big changes needed to make life great again. They know they were given the tools to have a great life and they know they were outsmarted. The great life they deserved was taken from them. They aren't sure who took it or how they did it, but they know it happened. My heart goes out to these men. I would like to see all people be satisfied with their lives. At the end of this article, I will explain who took their great life and how they got it.

I'm proud of my ability to solve puzzles. I am working on a solution for the unsatisfied lives issue. I believe I have some good concepts in the making. I'm afraid that my ideas may not be accepted by the people who need it the most because of traditions and beliefs. Some people have become so accustom to living around acceptable greed that they won't let go of their own greed long enough to enjoy life without it. I can only show one example. Several years ago, I decided jealousy was no good at all and that I didn't need it in my life. Last year, I realized that my greed was the one last thing that was making my life unsatisfactory. I let go of my greed and now I'm loving life.

So, who took their great lives and how did they do it? We already know why they did it. It's simple selfishness. It's easy to sit back while others work for you. The sad part is that the people who are taking the largest amounts are giving it to their children. This means that you are not working hard for the man, you are working hard for the man's children.

The great life was taken by lawyers who demanded more than what was deserved. This caused the general public to insist on everyone carrying large insurance policies. That caused several other groups of non-productive people to live off of your fruits. This problem can easily be solved by changing the law to make everyone have to carry their own insurance policy on themselves and they could only make claims against their own policy.

Another group of people who took the great life are the politicians who wouldn't work out their problems with other countries. These politicians took your fruits to make war with other countries. If an industry makes something that is not needed by our people (bombs and tanks) then it is classed as nonproductive. All of the people who work for that industry will live off of your fruits. I listened to many people say "Just bomb the hell out of them" when we were being attacked. How many of those people want to work the next ten years of their lives just to pay for those bombs, jets, and manpower needed to fuel the war they want? The economy seems to do better in times of war. This may be due to several factors, but any time we reduce the population, the remaining population gets to split up the property of the people who die. The number of houses remains the same. There are just fewer people to live in them.

I can make examples like these all day. The truth is that the working class is losing a little at a time to dozens of different places or people. Most of the fruits are being extorted by the smarter people who sit behind a desk all day and take 3 months vacation a year.

Also, as long as we keep paying too much to be dazzled, we keep paying another nonproductive group of people enough so they may live like royalty. This will pass. We will soon realize that these people are not as special as they seem. Who exactly? The circus owners, singers, actors, musicians, amusement park or carnival owners and anybody who gets paid way too much for the amount of work they do to dazzle us. Do you remember how crazy people would get for Elvis or the Beatles? As humans evolve past being star-struck, we will realize that they don't need to live so large.

Just recently, I watched a band on TV that was complaining about people pirating music from the internet. The musician sounded like he was about to cry when he said he wasn't making enough money. He needs to look at his producer. That is where most of the money is going. Learn to do what your producer does and you won't need them. Also, if music were priced right, people wouldn't feel so compelled to download copies from the net. We know that we are being charged too much for it. We pay $15 for a CD when it only costs about 0.50 cents to make. And worse yet, there are only 3 good songs on it. That's like paying $5 per song. I'm sorry, but I've got no sympathy for you.

THE CAR PUSHER

Back when I owned and ran a small automotive shop, I came up with a way for one person to push and steer a car with a dead motor. My shop was the highest point of the parking lot. When we had a dead car, we had a hard time pushing it up hill to get it in to the shop. I didn't always have 3 or 4 guys to help me push. A few times I was forced to work on the car out in the parking lot because I just couldn't get it inside on my own.

I designed a small plastic wedge that I could slip under the back of any tire on the car. This wedge had wheels on the bottom and top. It had two spring loaded arms that would wrap around the sides of the tire to keep it centered behind the tire even when turning. This wedge had a handheld drill motor and a rechargeable battery pack in it. I used a gear reduced type drill motor so that it would have enough torque. I connected a length of wire to the variable speed trigger from the drill so that I could sit in the car with the wire hanging out the window as I steered. It wasn't fast but it got the job done.

My first design had a flaw. The car would gain inertia while the motor was pushing. The pusher had no inertia and it would stop as soon as the trigger was released. The car would roll right out of it and cause me to have to get out and reset it to go again. My second design has a micro switch on the top side that would run the motor just enough to make the pusher keep up with the tire even when the trigger was out.

As a mechanic, I have pushed many, many dead cars. One trick I learned was that you could push the top of the tire to double your torque. This is because the wheel acts as lever in the same fashion that a chain hoist will multiply your lifting capacity. Technically, the force needed to push the car would be cut in half. This was only good to get the car started rolling. Once the car was at walking speed, the top of the tire was moving too quickly to keep up with it. This torque advantage is the reason why a hand held drill has the ability to push a 3000lb car. It doesn't go fast but it can allow one person to do the steering and push a car by themselves.

After designing this unit, I realized how simple, cheap and small it could be made. I didn't even need a battery pack on it. I already had wires going into the car for the trigger so why not install a cigarette lighter plug on them. This unit could be around the same cost as a decent handheld drill. It could be sold as part of a roadside kit to allow a stranded driver to get their car out of a travel lane or even finish the drive for that last ¼ mile to get to the gas station after you've run out of fuel. A larger more industrial unit could be sold to mechanic shops, if nothing else, to save them from having to take 3 or 4 other mechanics away from their work to help push the next dead car. Each mechanic would be able to do it by themselves.

MY GIFT TO YOU

As a thank you for purchasing my book, I will give you a gift that could be worth nothing or it could be worth a billion dollars and allow you to survive long after everyone else dies. It depends on how you use this gift.

I am going to give you the gift of foresight. I am going to tell you how the future will unfold and why. You will have the chance to use this information to make changes in your habits and collect the needed supplies in order live comfortably as changes happen to our world. Sorry, no lottery numbers!

What is going to happen and when; the first problem that we will deal with is the ocean rising. If you live on or near a beach, swamp, or river, your land will be covered in ocean water given enough time. How long we have before this happens is the controversial part. I have been predicting that we will see the first flood damage within 40 years. Some experts have agreed that we will see damage within the next 75 years but most of them have been asked to stay quiet about their findings.

The original 30 year old prediction was from 200 to 500 years for the flooding and other natural disasters to show there face. With more studies and research we had shortened our predicted time down to 100 to 200 years and this was the common belief for the last 10 years. Now, finally, some scientists are beginning to see what I see. Each natural disaster or event will cause damage that will accelerate the arrival of the next event to come. A good example would be how each hurricane we have will pull down a little more ozone and kick up a little more CO_2. This will damage our atmosphere to the point that we will collect and keep more heat from the sun each year that passes. This extra heat will cause more hurricanes the following year as well as melt more ice and raise our ocean level. Ice and snow reflects most of the sun's rays back into space. The more the ice melts, the less the suns rays will reflect into space. The extra sun on the surface means extra heat. Just recently, several experts have come to the conclusion that I've been saying for a while now. The first few global warming events will cause acceleration to the following events.

This is not the end of this vicious cycle. This extra heat will cause frozen lands in the artic regions to thaw. These lands are saturated in CO_2. This extra CO_2 will enter our atmosphere so fast that the next several years will get hotter very much faster.

It turns out that my prediction of seeing serious natural disasters within the next 40 years is probably the most accurate. I'm not saying that we are doomed in 40 years. I'm saying that we will have to move away from our lowlands and we might have continuing hurricanes and storms all year around. This will make it difficult to grow crops or build new construction.

Eventually as more and more ice melts, the ocean level will rise by approximately 240 feet. This may take as much as 150 years but don't underestimate how fast our climate is evolving. This means that it could happen as fast as 60 years from now. You may not be around to starve to death but your children may. If the storms make it difficult to grow crops, we will become dependant on living on fish. We will very quickly deplete our oceans of fish.

Once the ocean level rises, we will have to squeeze all living people on to the remaining land. I can't say with certainty, but I think that the surface of land above water will be reduced by about half.

My suggestion to you is to decide not to have children and convince others that the more people alive at the time this happens, the harder it will be to feed them. We need to begin reducing our population now. Please consider how much drama your children will have to deal with. Why would a person want to bring a child in to a world with these kinds of problems? Do you have to put them through that?

Another step you can make is to prepare now. Find out how many feet above sea level you live. Decide to move to an area at least 250 feet above sea level. The worst position to be in is to live 150 – 240 feet above sea level. All the people who live below 150 feet will move inland first. It will be much harder to find a good place to move if you wait too long. It may become difficult to build as we run low on construction materials and as the storms become relentless.

It may be difficult to buy some kinds of foods at supermarkets. You might want to start a garden of your own although you could have trouble growing foods once the storms become almost steady.

Any new construction should be built to handle 200 MPH winds even if it is built in the North West. The size and strength of the hurricanes to come is increasing. Hurricanes will travel inland much further each passing year until no land is safe. I wouldn't be surprised to hear that a hurricane will make it to Canada soon. We also have to consider that we will see superstorms that will travel in a repeating pattern, making loops around the country over and over, lasting 3 months at a time. We can expect this to happen around the same time our oceans level rises about 20 – 40 feet.

Eventually, cash and money of most forms will become worthless. It doesn't matter if you are super wealthy, who will sell you the last of their food or shelter if it means that they will die? Before this happens, the costs of food will sky-rocket. You can expect this to begin when the oceans level rises about 10 – 20 feet. You might want to stock up on foods that will last a while.

Over time, it will become increasingly more difficult to pump and transport fuels or wood for burning. It would be smart to collect tools and machines that do not need to be plugged in or buy a few of my incentive bikes so you will have a place to plug things in that doesn't require burning anything. Also, you would benefit from a few extra layers of insulation on your house once we reach a point where we will have nothing left to burn.

After the world floods, the temperature will begin to drop as the atmosphere changes. This is not going to be as easy to predict by scientists as other events of an ice age. It has been predicted that the planet will go into a deep freeze where the oceans will freeze as far south as possibly Georgia or even Florida. We are going off of the data collected about our last ice age and if it is anything like the last one, there will be no tropical zone. The weather in Mexico will resemble Canada's weather. This Ice age will be different. Our atmosphere will be significantly different this time because of human presence. We might not see a deep freeze this time.

If all unfolds as predicted then it turns out that this is the greatest time in our history to be alive. It will be 10,000 years before the planet will be back to the conditions similar to what they are now. It is possible that humanity may not survive through this ice age.

I feel like this is our last chance to try to stop this ice age from happening. I have to chuckle when I think about what we would have to do. Can you imagine what life will be like if we had to give up burning anything. The part that makes me chuckle is that I know that we are capable of living without burning anything but we are too stubborn to actually do it. We won't even give up cigarettes! Here comes the ice age!

WHY ARE WE CHARGING TO EDUCATE PEOPLE WHO WANT TO LEARN A SCIENCE?

I'm referring to the sciences that will help us figure out how to fix our global warming and health/disease related problems!

This should be free! These sciences are the key to our survival as a planet and as a race. We should only charge for teaching arts. Our government should foot the bill to educate those of us who want to see humanity live through this up coming ice age.

If our government doesn't help us get the education that we need, then we should do it ourselves. Who was it that said you have to have a degree of some sort to teach the knowledge you have? Any person who knows how to do something should take on apprentices. After work, a reasonable shift, an average Joe could go home and teach something he knows for a small and fair set amount of money. We can't let him set the price and say it's because he passed a test. We don't need to waist time with testing when our classrooms are small. As a teacher, you know who is learning and who isn't.

We should make it that you should always give the minimum wage in dollars for each lesson. If you feel that the teachings benefited you, then you buy them a gift. Of course it will be dependant on what you could afford.

The teachers will make a list of things that they would like as gifts. Half of the items on the list should pertain to learning or studying the subject and the other half should be for the basic needs or wants of the teacher. The students decide what they can afford. The gift will represent departing. It will say thank you for the lessons but this is my last one.

With today's technology and the internet, we can easily keep track of how each teacher or student compares. Each student will give a short review so that others may decide if they would like to join this class and each teacher will describe that student's enthusiasm toward learning or whether on not that student is suited for this area of study.

If you have a lot to teach or a subject in high demand, you could take on enough apprentices to support your self. In the event that an apprentice is not learning, you have to be willing to tell that apprentice that he needs to move on to something that better suites his potential.

A business owner could take apprentices to work with him to teach manufacturing systems or organization skills. We will not allow teachings that are related to earning money as this will not benefit our cause.

A Leader of Man

Today I caught myself singing a song in my head. The lyrics go as follows "I am not a leader of man... I prefer to follow." I like the beat so I repeated it several times. I noticed each time I got to the word "not" I would stumble. I realized at that point that I don't want to sing a song in my head if the lyrics don't apply to me. In this case, I like the song and I want to sing along but don't want to be heard saying something that is not true. I feel immoral when I lie. I think most people do.

If I would be lying by saying "I am not a leader of man.", then I feel solidly that I am a leader of man. Now, I realize this on a new level. The previous thought combined with this thought "A person must put themselves in the place where they want to be." equals my realization that my new goal is to put my self in a position of leadership.

I've always been working hard toward a goal. I wasn't sure what it was for a long time. I knew my parent's teaching wasn't without good intentions. I knew I wasn't striving toward better grades in school and raises at work just to go blend in with the blue collar class. I've worked hard in the blue collar class since I was 6 years old. I loved it so much. I had to beg my father to take me on the job. I would hurt myself on the job and my dad would stop taking me for a while after each time. I didn't want to miss a beat, so I learned to be safe and focused. I still love to work, when I can see my accomplishments. That is what I enjoy doing. But now I realize I am capable of teaching the world how to live much better lives through science. I don't even need to tell you that I am teaching you. I simply need to produce new inventions and tell everyone exactly how it works and how you can use it to make your life more comfortable.

There is no doubt that today we can use a good roll model and leader. It becomes my goal to put myself in a position where I will be capable of doing this. I don't want to lead through politics. I would rather lead by being recognized as an Einstein or Franklin first. I must begin distributing my inventions and methods until I am recognized like this. This is how I will win the attention of those who need a roll model, then I will lead by example. I am a blue collar worker. I work between 50 and 80 hours a week between my two jobs. I have two jobs so I can afford to provide a very comfortable life for myself and those closest to me. I am not afraid of hard work. How could I be a leader if I didn't work hard for the people I am leading? I feel that we need to work fast to stop injuring our environment as soon as possible. The sooner we make our worlds health a priority, the sooner we can relax in the thought that our children will have the chance to live much better lives then we did. Our own future comfort depends on what we do today. As I've stated before, we aren't going to colonize planets in other solar systems. We only have these few to choose from and Earth looks the best from where I'm standing. We need to make earth last. I can do that with your help. Am I a tree hugger? No. I believe we should use the resources around us to learn and develop methods of existing without injuring earth.

A FINAL NOTE

I need your support for this. The present laws may prohibit me from distributing certain inventions in order to protect the rights of a huge corporation to earn more money. I know I have to distribute my inventions regardless of the current laws. To hold back an invention that would provide betterment should be viewed as a crime against the well being of society. If I am faced with a situation where one of my inventions is taken or held back, I would appreciate your support to help overturn the law(s) that prevent its distribution. If you like my ideas in general and feel you would support my goals in any circumstance, please show me by logging on to www.inventionsbyfred.com and enter your information as a supporter. It's free. I just need to know that I have the support of the public when it's time to go head to head against huge corporations in court.

Many inventions are one minded. This means that a particular invention is only important to one person. It can be nearly impossible for that person to know what is important to the general public. When something is vital and indispensable to someone, they assume that other people will feel the same way.

The same applies to the opposite. If you see no need for something, you can't comprehend how others will support or care for it. This is how I feel about cigarettes, but that doesn't mean that millions of people won't feel the opposite.

Another thing to keep in mind is that it takes a tremendous amount of guts to propose, get behind, or support an invention. The only way any of these inventions is going to get off the ground is if the people who want to see it come to life join together to say so. Just knowing that you might purchase a certain invention is enough for me to show investors that there is a market for it. Please log on to my website, locate an invention that you like, and simply say "I would consider purchasing this invention." You may do this once per invention. Once enough people add their names to the list, we can give investors the confidence to invest money into something that they might not be sure about.

ABOUT THE AUTHOR

Fred Sanders is a puzzle solver. He uses his talents to make life more comfortable for all. His inventions speak volumes about his vision of a better world to come.